ARTIFICIAL EVOLUTION

How technology makes us think we're better than we are (and why that's dangerous)

ROB SNEDDON

Candlepin Press

ARTIFICIAL EVOLUTION

Rob Sneddon

ISBN-13: 9781672370776

Thanks to MerwinAsia for permission to quote the
passage from *Descent into Hell: Civilian Memories of the Battle
of Okinawa*, copyright 2014 by Ryukyu Shimpo, that appears
in Chapter Four

Cover and Book Design
Tammy Francoeur Sneddon
Cover Photo Illustration by Tammy Francoeur Sneddon
Background photo ©istockphoto.com/Tetiana Lazunova

Published by Candlepin Press
Somersworth, New Hampshire

First edition, December 2019
Printed in the United States of America

To Will, for helping me see the 21st century as it really is.

TABLE OF CONTENTS

Introduction vii

Chapter 1
Revolutions Per Minute... 1

Chapter 2
A School on the Move 15

Chapter 3
Good Forgetting vs. Bad Forgetting 41

Chapter 4
Nuclear Weapons Don't Kill People, People Kill People 57

Chapter 5
Not Intended for Outdoor Use............................. 87

Chapter 6
Ask Your Doctor if Immortality Is Right for You 103

Chapter 7
Recalculating... 117

INTRODUCTION

"Don't let me be bound hand and soul to the telephone."
—Alexander Graham Bell

I have an old rotary phone that I keep in a drawer. It belonged to my late parents. Their number, which I can recite from memory, is embossed on the clear plastic disc in the center of the dial.

I hang onto that phone mostly out of sentiment. It's both a personal and a cultural artifact. But I also keep it for another reason. It still has a practical application. I realized this a few years ago during a power failure. Landline service was available, but we lost it at our house because the base on our cordless phone needed power.

I had an idea. Since a rotary phone required no external power source, maybe my parents' old phone would work as an emergency replacement.

It did. So while the power was out, my wife and I did something we hadn't done in decades. We dialed our calls.

~

Whenever we cross a technology threshold, we leave something of value behind. And we often leave ourselves a little more vulnerable. In *Isaac's Storm*, Erik Larson notes that when lightning knocked the Galveston power station offline a few days before the city's most devastating hurricane, residents struggled to find emergency light sources. Even the cops were caught short; there wasn't a single candle at the police station. The entire island had gone all-in on newfangled electric lights.

That was in 1900.

Since then all of America has become much more dependent on technology—a dependency that becomes apparent as soon as we're deprived of that technology. Whenever I get booted off the grid I feel a low-grade loose-endedness. With no lights, no computer, no TV, no microwave, no electric stove, no electric *anything*, how am I supposed to get through the day? When the power goes out, I just kill time until things get back to normal.

The unsettling thing is that what's normal in 21st century America is not normal in the context of human evolution. According to the book *Whiplash: How to Survive Our Faster Future*, if the time from when hominids first started walking on two feet until the present were compressed into a single year, the industrial age wouldn't begin until just before dawn on December 31.

That means the power grid wouldn't arrive until sometime that afternoon, less than 12 hours before the end of the year. In other words, only about one-seventh of 1% of human history has occurred during what we could roughly call modern times.

Life is so different from what it was a few thousand years ago—or even a few hundred years ago—it's as if we're living on a different planet than our forebears did. I can't imagine going through an entire lifetime without ever flicking on a light. But billions upon billions of people managed to do it. And in some respects they were better grounded than those of us who have lived our whole lives inside the technology dome.

I'm not saying that life was necessarily better back before the world powered up. I'm just saying that every new upgrade in technology comes with a price. And it's not always immediately apparent what that price is. When people first started burning fossil fuels, they weren't worried about melting the polar ice caps.

Sometimes the toll is so insidious that we barely notice it. Like the hidden cost of smartphone technology. I'm not referring to the obvious dangers associated with smartphones, like the estimated 11 teenagers a day who die as a result of texting while driving, or the almost unimaginable toxicity that results from processing rare-earth metals in China. I'm talking about what we've sacrificed in the name of convenience.

I was born in 1959. The telephone has evolved exponentially in my lifetime. When my wife and I used our emergency rotary phone, we laughed at how cumbersome it was. It took about 12 seconds to dial a seven-digit number. If that doesn't sound like an especially long time—well, just pause for a moment and count it out.

Now imagine it took that long to place every call. (There's no way to program numbers on a rotary phone.) Also bear in mind that, for much of the rotary era, both you and the person you hoped to reach had to be at a fixed location, with access to a landline, at the same time. If you didn't reach them, there was no way to leave a message.

Smartphones let you communicate with just about anyone from just about anywhere at just about any time, almost instantaneously. Change of plans? Send a text. Forget what you were supposed to pick up at the store? Just tap your home number, or say your spouse's name, and you can find out while you're still in the produce aisle. Need to get a detailed message to someone at work after hours? Leave a voicemail. And that just covers basic telephone functions, without getting into all the apps that have transformed phones into tiny portable PCs.

In terms of efficiency, rotary landline phones were about four score and seven years behind smartphones. There's simply no arguing that point.

I'm going to argue it anyway.

The old phones enforced a different kind of efficiency, born of simplicity. We never gave them any thought. When my parents bought their first house, in 1961, this was their phone plan: They had a basic black Bell Telephone desktop model installed in the front hall. And, as far as I remember, they didn't have to do anything else with their phone during the 18 years that they owned the house. They didn't have to figure out how to use it. They never lost it. They never broke it. They never had to charge it. They never had to replace the battery. They never had to replace the entire phone, even though it needed only a battery, because the battery was obsolete. (I had to do that with a cellphone one time.) They never worried about anyone stealing it. They never lost all their contact information. They never spent an evening comparing various service plans because an endless stream of TV commercials made them feel like their current provider was playing them for suckers. (A phone was just a phone—not the key to a better life.) They never discovered that their phone was incompatible with some fabulous new app that all their friends had. In fact, during the whole time we lived in that house, our phone never had any apps at all, not even an answering machine. (Unless you count my Scottish grandfather. One time, during an extended visit to the U.S., he was alone at our house when the phone rang. He picked it up and—as we later learned from the perplexed caller—blurted, "There's nob'dy here!" then hung up. Wish I had a recording of that; I would make it my voicemail greeting.)

For all the obvious inconveniences, limited phone service forced people to manage their lives better. They had to consider other people's time, not just their own, when making plans— and those plans had to extend more than a few minutes into the future. Since the advent of the smartphone, a disturbing number of people think the capability to let others know they're running late makes it acceptable to *be* late. ("I sent you a text. Didn't you get it?")

In addition to transforming America into a procrastinator's paradise, constant connectedness has deluded people into believing the "efficiency fallacy." Overscheduled executives think they're maximizing productivity by making business calls from their cars, for example. Never mind that the people on the other end of those calls often end up with incomplete or inaccurate information because they either misheard the person or couldn't hear them at all. And there's a good chance executives won't remember much about calls they made from the car anyway. They can't document anything while driving and, in my experience, they often sound distracted—which is probably a good thing. Driving a car is not just mindless downtime. It's actually a pretty demanding activity. It's not something you should do while discussing matters of any consequence over the phone.

Phone-related preoccupations have led to all kinds of mishaps and mistakes that we never used to see before. Whenever there's a major accident involving public transportation, the first thought that pops into people's heads is, *Was the driver texting?* And remember Warren Beatty announcing the wrong winner for Best Picture because the accountant in charge of the Academy Award envelopes lost track while tweeting during the show? Or the false alert of an incoming ballistic missile that triggered a panic when it pinged across Hawaii? Neither of those things would have happened in the rotary-phone era.

~

Phones have been a distracting influence since day one. (Alexander Graham Bell's faithful assistant, Watson, achieved immortality by becoming the first person whose work was interrupted by a phone call.) What's different now is that social media apps and other accessories have introduced so many new ways for phones to erode our ever-shrinking attention spans. (Here's depressing evidence of how widespread smartphone addiction has become: In a study, 37.4% of the participants essentially admitted that they valued their phones as much as they did their close friends, if not more so. And a recent *New York Times* article, echoing my thoughts about rotary phones, carried the headline, Is the Answer to Phone Addiction a Worse Phone?)

Too many people allow technology to control them, rather than the other way around. They do this, I think, in the unconscious (and misguided) belief that technology provides a shortcut to evolutionary advancement—to superior-beinghood. For a species as complex as ours, genuine evolution should take about a million years. But in just the past 200 years we've hacked the process so that even the slowest and stupidest among us can now move faster than a cheetah, fly higher than an eagle, and routinely perform miracles of science that would have dazzled Einstein.

And once you've had a taste of that, you want more. Technology overcomes obstacles. It makes life easy.

Until the computer, the car, or the plane crashes. Then we're rudely reminded that we're no more advanced, on an evolutionary scale, than the Pilgrims on the *Mayflower* were. In fact, in a Darwinian sense, we might be worse off than Myles Standish & Co. With little technology to speak of, the Pilgrims managed to start a new society in a strange, hostile environment with no infrastructure or support systems.

How many of us could do the same?

And yet we regard our deteriorating self-sufficiency with First World smugness. We don't worry about elemental survival skills because we don't need them anymore. It's as if each incremental gain in comfort, convenience, and security has not only made our lives better—but it has also made *us* better.

Watch a couple of hours of random American TV and you'll see this attitude reinforced at every commercial break. The essential message, over and over, is: *Here's a way for you to save even more time and make even less effort in your daily life!* Often, the subtext is: *And if you* don't *do this, you're a jamoke who can't keep pace with modern America and will be condemned to a miserable, lonely, ostracized life.*

In 2016 DIRECTV ran an ad campaign called "The Settlers" that perfectly captured this glib sense of superiority. This was the premise: Anybody who settled for mere cable service, instead of DIRECTV, was as hopelessly behind the times as 19th century homesteaders. Might as well make your own clothes or hunt your own food—as if the ability to do those things is worthy of ridicule. Settlers, as portrayed by DIRECTV, were just the butt end of the yardstick we use to measure how much better life is in the 21st century than it was in the 19th.

Yeah—it's just a joke. I get it. The premise is a clever play on words, and the commercials are well executed, with sophisticated production values. Still, the underlying attitude betrays "a form of bigotry directed at the past," to steal a phrase from the writer Bill James.

It is also a form of bigotry directed at a certain class of people in the present. Not everyone can keep up with the latest technology—financially or otherwise. And not everyone wants to. But it's increasingly obvious that technological proficiency is the primary thing that separates the haves from the have-nots in 21st century America.

Brink Lindsey, a senior fellow at a libertarian think tank in Washington, explores this idea in a book called *Human Capitalism*. "The less adept you are at coping with complexity," Lindsey writes, "the humbler your position in the social hierarchy is likely to be." He concludes that "fluency in abstraction" is the primary determinant of success in modern America.

I agree with Lindsey's point—to a point. I understand the need to embrace new technology as the demands of your life and your career change. I wrote my first published article on a typewriter, using notes I'd scribbled in longhand. I mailed the manuscript to a magazine in a manila envelope. I have no desire to go back to those days. And I felt no need to hang onto my old Smith Corona, the way I did my parents' old rotary phone.

The reason I regard those two situations differently comes down to fundamental utility. Strip a smartphone of its accessories, and its basic purpose is the same as a rotary telephone's: to enable you to converse in real time with somebody at a different location. In fact, the rotary phone was better suited for this purpose in some ways. The handset was more ergonomic; it was designed to fit your face, not fit in your pocket. Also—and this is no small thing—the sound was clearer.

As a writing instrument, on the other hand, a laptop is vastly superior to a typewriter. "Writing" is really just a process of organizing your thoughts. A laptop makes it much easier to edit and revise your thoughts in real time than a typewriter ever could.

So while the devices I use in each case have evolved drastically in my lifetime, talking on the phone still feels the same to me as it ever did, but writing feels like a different—and much improved—process.

~

In sum: This book is not an argument against progress. It's an argument against relentless innovation for its own sake—a process that threatens to reduce us to bystanders in our own lives. "As task after task becomes easier, the growing expectation of convenience exerts a pressure on everything else to be easy or get left behind," Tim Wu wrote in a *New York Times* essay called "The Tyranny of Convenience." "We are spoiled by immediacy and become annoyed by tasks that remain at the old level of effort and time."

And we are all complicit in allowing this to happen. "Public consciousness has not yet assimilated the point that technology is ideology," Neil Postman wrote in *Amusing Ourselves to Death: Public Discourse in the Age of Show Business*. "This, in spite of the fact that before our very eyes technology has altered every aspect of life in America during the past 80 years."

Postman wrote that in 1985, before the internet wrought

yet another fundamental change in American life. And yet we *still* don't get it. But whether we recognize that technology has become our ideology or not, Postman added, "all that is required to make it stick is a population that devoutly believes in the inevitability of progress."

While I think that progress is possible, and even likely in the short term, I don't think it's inevitable in the long term. In fact, I don't believe that America's current trajectory is sustainable. I fear that we face a terrible reckoning for surrendering so thoroughly to the lure of technology. The world's population has more than doubled in my lifetime, from about three billion to 7.6 billion, and is expected to hit 10 billion around the middle of this century. That's also right around the time that many scientists believe global warming will reach a dangerous tipping point. If both projections come to pass, then by 2050 there will be a lot more people competing for fewer and fewer resources on an increasingly toxic, unstable planet.

Nor do I have much faith that we can prepare for this eventuality simply by stepping up the pace of artificial evolution and developing still *more* technology. "It's a very dangerous game," Stanford earth scientist Rob Jackson, chair of the Global Carbon Project, told *The Atlantic*. He noted that the most optimistic long-term projections for mitigating the manmade effects of climate change are predicated on devising some form of large-scale carbon-scrubbing technology. "We're assuming that this thing we can't do today will somehow be possible and [affordable] in the future," Jackson said. "I believe in tech, but I don't believe in magic."

Neither do I. That's where I part ways with Brink Lindsey and other writers who conclude that the key to survival in the coming years is to simply ride the wave of innovation, push the envelope, and stay ahead of the curve. "If you want to raise a middle-class, college-educated knowledge worker," Lindsey writes, "you need to get started as soon as possible, and you need to keep at it unrelentingly."

Good advice for now, I suppose. But when the social fabric starts to tear, when food and water shortages and massive exoduses of refugees become widespread, when mutant viruses and antibiotic-resistant strains of bacteria create unchecked pandemics, I'm not sure how much help a middle-class, college-educated knowledge worker raised unrelentingly to attain fluency in abstraction will be. It probably wouldn't hurt to make friends with a few families like the Settlers, just in case.

CHAPTER ONE

Revolutions Per Minute

"Hello, we're glad you made it! Welcome to the future!"
—The Firesign Theatre

One of my favorite pieces of reality filmmaking is a 12-minute short called *A Trip Down Market Street*. It was produced in 1906 by some bicoastal filmmakers, the Miles brothers, using a camera mounted on the front of a San Francisco cable car. The film gained widespread popularity after a *60 Minutes* segment revealed that it had been shot just days before the great San Francisco earthquake.

Knowing what's about to happen to the people in *A Trip Down Market Street* gives the film undeniable power. But there's another reason I found the timing fascinating. The film also captured the chaos of progress in progress.

We tend to think of progress as linear and discrete, like a "stages of man" exhibit. The vinyl record begat the cassette tape (along with a genetic mutation called the eight-track) which begat the CD which begat the MP3 file. But technological innovation rarely proceeds in neat, measured steps. There's overlap and backlash. Some people still prefer vinyl.

Watching *A Trip Down Market Street* is like watching a wrestling match between the 19th century and the 20th. Cable cars and streetcars share the road with horse-drawn carriages and automobiles, along with bicycles, newsboys, pedestrians (who are struggling to keep up with the traffic) and street sweepers (who are struggling to keep up with the horseshit). They all weave among each other without a single traffic signal to regulate the flow. The chaos is vibrant.

Granted, some of it is also contrived. The same three automobiles keep driving around the cable car, making it appear is if there's much more vehicular traffic than there was. Still, automobiles running circles around every other form of transportation is an apt metaphor for the dawning century. In more ways than one, the people in *A Trip Down Market Street* didn't know what was about to hit them.

~

Manmade transportation got off to an excruciatingly slow start. Circumstantial evidence suggests that Australia's first inhabitants reached the continent by using rafts about 50,000 years ago. For roughly the next 49,800 years there was little improvement in water travel. Every boat during that time was propelled either by oars or sails or prevailing currents. Those restrictions imposed a natural speed limit that was unchanged for millennia. (The lack of reliable navigation aids further limited water travel.) It took the *Mayflower* about the same amount of time to cross the Atlantic in 1620 as it had taken the *Nina*, the *Pinta*, and the *Santa Maria* in 1492. To put that in perspective: Imagine not being able to travel any faster today than people did in the 1890s.

Things progressed no faster on land. For thousands of years, people got around mostly by foot or occasionally on horseback. It wasn't until inventors harnessed the power of the steam engine in the 19th century that things changed. And they changed quickly, on both water and land. The Erie Canal opened in 1825; America's first railroad, the Baltimore & Ohio, began a year later. That's when the Industrial Revolution truly began to gain traction in the United States.

What had been a methodical, two-pronged attack on the limits of mankind's locomotion became more intense and, paradoxically, more diffuse as the 19th century drew to a close. Some inventors set their sights on conquering the last remaining transportation frontier, the air. At the same time, others sought to adapt the technology of rapid ground transportation for individual use. That would grant Americans unprecedented independence to choose where and how to live.

That simultaneity expanded humanity's reach with astonishing speed. The airplane was invented at roughly the same historical moment as the automobile—a moment when still other inventors focused on perfecting another newfangled transportation device called the bicycle.

This led to odd juxtapositions and strange bedfellows. The Wright Brothers developed the airplane in their Ohio bicycle shop. Another bicycle shop owner, Carl Fisher, dreamed up the Indianapolis Motor Speedway. The speedway's first race, in 1909, featured neither bicycles nor cars but helium balloons.

Americans were obsessed not only with how fast people could go but also how far. Ray Harroun, winner of the first Indianapolis 500-mile automobile race in 1911, covered that immense (for the time) distance in less than seven hours. Later that year, Roald Amundsen led the first expedition to the South Pole—two years after Robert Peary famously (if dubiously) claimed to have been the first to lead an expedition to the North Pole. Explorers had literally gone to the ends of the Earth.

In *River of Shadows*, Rebecca Solnit notes that "annihilation of time and space" became a stock phrase in the 19th century. The manifold methods that inventors and explorers used to commit that annihilation extended beyond planes, trains, and automobiles. Electric lights turned night to day. The telephone rendered distance irrelevant when communicating by spoken word. Photography—the subject of Solnit's excellent book—froze the present.

There were no controlled studies to see how all these rapid developments would impact the human psyche. They just *happened*. Which meant that after living basically the same way for thousands of years, people had to learn to live an entirely different way with little warning. Change became the rule, not the exception. How did that affect people's minds—particularly among those whose lives began during one era and ended in another? There's no way to say for certain, but it's tempting to speculate.

Consider the life of an obscure explorer named Delia Akeley, who died in Florida in 1970. She was born in Wisconsin in either 1869 or 1875, depending on which sources you trust. Either way, she entered a world without telephones, electric lights, automobiles, or airplanes and left one in which people were walking on the moon and nuclear Armageddon was a legitimate threat. That's a lot of technological change to process. Maybe it's not a coincidence that she spent a good bit of her life exploring remote parts of Africa.

~

It wasn't just people who were subjected to rapid change. The world itself transformed as it was retrofitted to accommodate new technology. (There's no debate among scientists over whether the Anthropocene has begun; the only question is exactly when it started.) For telephones and electric lights to be practical, for instance, wires had to connect not just every city in the United States but also every *house*.

Today we take that cat's cradle of overhead wires for granted. But think of how crazy that idea must have seemed when it was first proposed. (And it won't be long before it seems crazy in hindsight, too. Once all cables are buried underground—or wireless technology develops to the point where cables aren't needed at all—people will find it jarring to look at all those ugly wires in movies and photos from this period.)

Also, cars could go only as far as roads would take them—which wasn't very far at first. A nationwide network of paved streets and highways was as critical to the development of modern life as the automobile. Try to imaging getting around 21st century America without the 47,865-mile Dwight D. Eisenhower National System of Interstate and Defense Highways.

~

Eisenhower first developed an interest in improving the nation's roads in 1919. As a lieutenant colonel in the U.S. Army, he accompanied a military convoy that crossed the country as a feasibility study/publicity stunt. The trip took 62 days.

Eisenhower wasn't impressed. To him, the exercise proved that the U.S. needed a much better transportation infrastructure. At the time, he was thinking purely in terms of conventional roads. But later, as Supreme Commander of the Allied Expeditionary Forces in Europe during World War II, he was struck by the efficiency of Germany's autobahn, the prototype four-lane superhighway. "The old convoy had started me thinking about good, two-lane highways," he wrote years later, "but Germany had made me see the wisdom of broader ribbons across the land. This was one of the things that I felt deeply about, and I made a personal and absolute decision to see that the nation would benefit by it."

As Earl Swift points out in *The Big Roads*, Eisenhower gets too much credit for developing the Interstate Highway System. No one man could have been solely responsible for such a massive undertaking; much of the groundwork had been laid years before Eisenhower took office. Nevertheless, the Interstate Highway System was one of Eisenhower's top priorities when he was elected President in 1952.

That year saw another transportation milestone, although you are probably unaware of it. It's also highly probable that you never heard of the aviation pioneer who accomplished it, even though his feat was more relevant to your life than the feats of Charles Lindbergh, Amelia Earhart, Chuck Yeager, or Neil Armstrong.

On May 2, 1952, Alastair Majendie piloted the world's first commercial jet. The BOAC Comet, with 36 passengers, flew from London to Rome—from the seat of one faded empire to the seat of another—in 2 hours and 46 minutes. It was the first leg of a 6,724-mile journey to Johannesburg that would take almost 24 hours and require three separate crews. (The same flight, by conventional propeller aircraft, took *10½ days* just 20 years earlier.)

The unprecedented flight set an immediate precedent for commercial jet travel: It arrived 19 minutes late for its first stop. (Majendie called this "rather disappointing.") By the time the flight reached Johannesburg, however, it had made up the deficit and then some. BOAC officials boasted that the flight proved Britain had at least a four-year head start on the United States in commercial jet travel.

BOAC was right. Commercial jetliners didn't begin service in the U.S. until 1958. And this evolutionary step was not universally applauded. "It was to have been comparable in enthusiasm, joy, excitement, and significance to Robert Fulton's steamboat ride on the Hudson River," *The New York Times* reported, "for the jet will do for commercial aviation what the steamboat did for passenger shipping. But to the dismay of almost everyone [involved] in and interested in aviation, this country did not soar as boldly into the wild blue of the jet age as most people had expected and hoped."

Complaints about the noise near Idlewild Airport (now called John F. Kennedy International) led to New York's "stumbling through a fog of acrimony, charges, insinuations and public relations releases, and backing into the day of the big jets. It was reminiscent of the early days of the automobile when the derisive cry was 'Get a horse!' "

But, just as the automobile quickly overcame the initial backlash and supplanted the horse as the average American's preferred mode of personal transportation, jet flight soon gained widespread acceptance as America's choice for long-distance travel.

My personal introduction to the jet age occurred on May 4, 1971, when my mother took my younger brother and me to Scotland to meet her parents for the first time. I was just shy of 12 years old.

Like the very first jet passengers almost 20 years earlier, we flew BOAC. (That was an acronym for British Overseas Airways Corporation, the forerunner of British Airways.) Our flight went nonstop from Toronto International Airport to Prestwick Airport near Glasgow.

I spent the flight enveloped in a sense of wonder. Even though jet travel was unlike any experience I'd ever had, I wasn't scared or nervous about it at all. I had an unquestioning faith in the integrity of the airplane and confidence in the people flying it. The moment I remember most vividly is looking down at the southern tip of Greenland, a frozen Arctic wasteland seven miles below. I couldn't have been more enthralled if I had flown to the North Pole on Santa's sleigh.

~

These days I'm more likely to feel a sense of horror over flying than the childlike sense of wonder I felt the first time. For a few years I had to fly fairly often for work, and I started having intermittent panic attacks. Once, just before takeoff, I unbuckled my seatbelt and almost bolted for the door. Only my even greater fear of causing an embarrassing scene (and possibly getting arrested) kept me from fleeing.

Here's the thing that people need to understand about the fear of flying, at least in my case. It really is a fear of *flying*— not a fear of crashing. While awaiting departure, I become hyperaware that what is about to happen is not a remotely normal experience for a human being. I have the heart-pounding realization that I'm making an absurd-sounding leap of faith— believing that this big metal tube can defy gravity and fly 500 miles an hour, 35,000 feet above the earth.

Pondering the sheer Buck Rogers implausibility of that notion can loosen my hold on sanity. *Really? We're all going to just accept that being catapulted a thousand miles through the troposphere is not only possible but also so routine that we don't even need to wear special equipment?*

During the period when my fear of flying was at its worst I confronted a profound dilemma. At the time I was an assistant editor for a pair of auto racing magazines. Writing about auto racing might seem an odd career choice for someone given to questioning our society's dependence on technology, but it made a twisted kind of sense. Success in auto racing required the driver to balance a machine on the absolute edge of control for an extended time. It could be breathtaking to watch. But when a driver, or drivers, went beyond the edge of control, that breathtaking balancing act came to a sudden, violent end.

To stretch a metaphor to the breaking point, I now feel the same way when I observe the human race as I used to feel when watching an auto race. For now it's a dazzling spectacle of speed and noise and technical proficiency. But I fear that a hideous accident could bring everything to a frightful halt at any moment.

Anyway, back to my dilemma. In 1995 an aviation enthusiast named Donald Pevsner arranged an attempt to break the world record for circumnavigating the Earth in a passenger plane. I say "arranged" because Pevsner wasn't a pilot who would actually make the attempt himself; he would simply be along for the ride on a chartered Air France Concorde. To foot the bill, Pevsner got Coors Light to sponsor the flight. At the time, Coors Light also sponsored NASCAR driver Kyle Petty, so he would be along for the ride, too.

Pevsner invited members of the press to come along and document the flight. Because I worked at a magazine that covered NASCAR, one of those invitations landed in my inbox. Would I like to accompany Kyle Petty on an around-the-world flight?

What to do? On the one hand, this experience would amplify the things I dreaded most about air travel. The Concorde would fly much faster (almost 1,350 mph), much higher (almost 60,000 feet), and much longer (about 31½ hours) than any other flight I had ever been on. It would require seven takeoffs and (one hoped) an equivalent number of landings— all without the one benefit that made me endure flying in the first place: the opportunity to get where I wanted to go in the quickest, most practical way possible. If all went according to plan, Flight AF 1995 would end up back where it started: New York.

On the other hand: How many people can say they not only went around the world but also did so in record time?

I decided to go.

I called the resulting article "Going Nowhere Fast." Despite the exotic itinerary—New York to Toulouse to Dubai to Bangkok to Guam to Honolulu to Acapulco and back to New York—the flight had "all the glamour of a bus ride to St. Louis," I wrote. We never ventured much beyond the arrival/departure gate at any of the airports we stopped at, and we barely had time to stretch our legs during refueling. Before our initial takeoff, Pevsner had issued the passengers a warning: "Anybody who's in the gift shop in Bangkok looking at an enamel elephant when it's time to go is gonna get an awful lot of Thailand for their money."

Notwithstanding the long, tedious nature of the trip, I found it less stressful than a typical two-hour domestic hop from, say, Boston to Cleveland. As I said, the routine nature of a "routine" flight is the very thing that I find unnerving. Flight AF 1995, by contrast, was conceived as a unique venture, and people approached it that way. We were a planeload of Phileas Foggs—only instead of going around the world in 80 days, we were going around the world faster than you could say Jack Robinson. Or at least faster than you could *hear* someone say Jack Robinson, since the flight was supersonic. (As I noted in the article, the mind takes an odd turn when deprived of sleep and confronted with abstract concepts like the International Date Line. If memory serves, we left Guam on Wednesday and arrived in Honolulu on Tuesday before rejoining Wednesday en route to Acapulco.) I embraced the jetlag and disorientation. Not knowing exactly where I was, what time it was, or even what day it was from one moment to the next was oddly exhilarating.

Flight AF 1995 did indeed return to New York in record time. Being along for the ride felt surreal in a good way, not a frightening way. It was like being 12 again. Soaring along at speeds approaching Mach 2, almost 12 miles up—so high that the sky started looking more black than blue—felt like a rare privilege.

~

Turns out it *was* a rare privilege. Flight AF 1995's circumnavigation record still stands, and both British Airways and Air France retired their commercial Concorde fleets in 2003. Thus ended an era of supersonic passenger service. It had begun in 1969, the same year Neil Armstrong became the first man to walk on the moon.

I have a 12-year-old son. I find it odd that two things that happened during my lifetime—people walking on the moon and passenger jets flying faster than the speed of sound—have yet to happen during *his* lifetime. (A Colorado company called Boom hopes to have a new line of supersonic passenger jets available by 2023.)

Despite the accelerating rate of change, and people's eager acceptance of any technology that makes life easier or more convenient, some technological advances are still ahead of their time—because they're beyond most people's means. List price for a round-trip Concorde ticket between New York and London was around $10,000 in 2000. Economics is, and has always been, one of the two main drivers of evolving technology. (War, as we will see later, is the other one.) The automobile didn't truly catch on until Henry Ford introduced the Model T, which was both mass-produced and mass-consumed. The price actually dropped from $850 when it was introduced in 1908 to less than $300 in 1925.

Once a technology becomes affordable enough to advance beyond the early adopters and gain widespread acceptance, it becomes entrenched. At that point it becomes difficult, if not impossible, for people to conceive of giving it up. Convenience is addictive.

For the average traveler, flying at 500 mph has remained the benchmark since the 1950s. Very few people got to experience flying 1,340 mph on the Concorde because very few people could afford it. So very few miss it. And that's probably just as well. Can you imagine what would happen if Americans were told that instead of driving 70 mph on the interstate, they could only drive 25 mph from now on? That's basically the step-back from flying on the Concorde to flying on a conventional passenger jet.

~

There was a point in the fairly recent past when Americans *were* told they would have to slow down their driving to what now sounds like an absurd degree. During World War II, the federal government imposed a nationwide speed limit of 35 mph in an effort to conserve rubber. (Tires wore out a lot faster back then, and the reduced speed was supposed to make them last longer.) And it's not as if cars of that day were markedly slower than today's. A 1941 Buick Century could easily cruise at 70 mph.

So was there an outcry over this severe cramp in America's style? No. A Gallup Poll found that 87% of America's car owners supported the measure. The 35 mph speed limit remained in effect throughout the country from 1942 until the war ended in 1945.

Picture the reaction if the federal government were to try to impose that draconian measure now. Granted, it's hard to imagine a scenario where an attempt to enforce a 35 mph nationwide speed limit would be even remotely plausible today. So, for the sake of argument, imagine a ridiculously implausible one. Say there was a burst of weird sunspot activity that produced intermittent, unpredictable visual distortions among drivers who exceeded 35 mph. If you drove beyond that limit you would probably be fine—but there was a 15% chance that at any time you could experience a sudden onset of extreme vertigo that, in all likelihood, would cause you to crash.

Now imagine that, as a safety measure, the government imposed a federal speed limit of 35 mph for as long as the phenomenon lasted, which experts estimated could be up to three years. How many Americans do you think would die in sunspot-related crashes during that time?

I'm guessing the number would be in the tens of thousands, if not higher. Absent a tangible hazard like snow-covered roads, a 21st century American wouldn't be able to stand driving on the interstate at a maximum speed of 35 mph for any length of time. In the hypothetical sunspot scenario, most people would soon go back to driving the same highway speed as before and just hope to get away with it—sort of the way people just hope we can survive global warming. Making such a drastic change is too hard.

Telling people that Americans collectively agreed to slow down to 35 mph in World War II wouldn't shame anyone into complying with the new sunspot speed limit, either. Violators would offer a version of this rationalization: *Well, life is much faster-paced today than it was back in the 1940s. People didn't have as much going on back then.*

Other than trying to win the most consequential war of the 20th century, if not of all time, of course.

Ken Ilgunas wrote about this numb denial in *Trespassing Across America*, his account of hiking the length of the proposed Keystone XL Pipeline. The pipeline would connect the tar sands of Canada to a refinery on the Gulf Coast of Texas, extending like a hypodermic needle through America's heartland to provide another fix of fossil fuels. And yet few people Ilgunas encountered along the way seemed to care. "The last thing we need is more oil," he wrote. "We need, rather, different consumption habits, a whole new relationship with the world. We need to quit destroying everything out of a sense of 'need,' when all we really need is a fucking sweater."

CHAPTER TWO

A School on the Move

*"You and I come by road or rail,
but economists travel on infrastructure."*
—Margaret Thatcher

Fear of flying is the most extreme manifestation of my technology-induced vertigo. But I feel milder strains at other times.

I'm especially susceptible in heavy urban traffic. I was 16 when the weirdness descended over me for the first time. My dad was driving my two brothers and me to a lake in Canada for a weeklong fishing trip. We lived in a rural area of western New York and were headed to an even more rural area of southern Ontario. But to get there we had to go through Toronto.

I had been in big cities before, including Toronto, without feeling any anxiety. And nothing out of the ordinary happened as we passed through Toronto that day. We didn't witness a fatal accident or get stuck in an epic traffic jam. We were just cruising through Canada's largest city in a red 1974 AMC Matador, listening to the radio hits of the summer of '75, like "Love Will Keep Us Together" and "Rhinestone Cowboy."

I felt torpid amid the superhighway rush, taking in the passing wash of asphalt and Jersey barriers and light stanchions and overpasses and enormous signs and the towering downtown skyline.

And then all at once, as I fixated on a crane on top of a skyscraper, I felt a darkening sensation, as if a cloud had passed in front of the sun. The tableau outside my window turned strange and mirage-like. It hit me that nothing in my field of vision was organic. *Nothing.* Human hands had created every single thing I could see and placed it exactly where it was.

That shift in perspective unsettled me. For a moment the name *Toronto* sounded like gibberish in my mind. And as a concept "Toronto" seemed like a contrived reality—which it was. That realization left me feeling disconnected. "Civilization" suddenly felt like the equivalent of a false front on a Hollywood lot, an elaborate attempt to make everyone forget that we're all just animals living on a planet made of dirt and water and sky. The sensation was fleeting but vivid.

I've felt that disorienting sense of having blundered into some alternative, artificial world a few more times over the years. When I was in my 20s I tried to capture the sensation in some song lyrics. I still remember them. (Just be grateful you can't hear me sing them.)

A School on the Move

Driving along on the fringe of the city
A little part of the evening highway
It's a big moving picture of tiny details
Change one and the whole thing derails
And trying to stay on top of it all
Is like counting guppies in a goldfish bowl
Follow the leader, the man in the street
Try not to lose that beat
Try not to lose that groove

Try to stay with a school on the move

Hard to believe it's all man-made
Hard to believe it's all man-made

Like a glimpse of fish in sunny water
The setting sun glints on a fender
Electric night at the end of the day
The Third World is two worlds away
They know nothing of our breakneck pace
They know nothing—is that a disgrace?
They know nothing, but what can we do?
We've gotta learn fast when the school's on the move

Hard to believe it's all one place
Hard to believe it's all one place

I'm aware of the risk in recounting my firsthand experiences with panic attacks and existential vertigo. It would be easy to dismiss me as unhinged.

In truth, those episodes are rare and they pass quickly. Like everyone else, I spend most of my time preoccupied with the minutiae of day-to-day life. I'm so busy keeping afloat in a flood of information and trying to dodge a procession of choices that I don't want to make that I have little time for big-picture contemplation. Even as I write this book, my concentration splinters. Messages keep bubbling up my on laptop asking if I want to update something. *No, I don't! Can't you see I'm busy? I like everything the way it is and just want it to stay like this until I can complete one measly thought. Is that too much to ask?*

Ubiquitous technology dictates the ground rules and pace of modern life whether we like it or not. A basic level of proficiency is the price of participation. And because people are both social and tribal by nature, they *want* to participate. (This is most obvious among school-age kids, with their naked desire for the newest phones and game systems.) If you are born and raised in the United States and just want a normal life, you *have* to adapt to technology and its complex infrastructure. It's hard to function in 21st century America without a car and a phone and a computer and a collection of passwords and PINs—without allowing data and algorithms to define to a large extent who you are.

Managing the perpetual upgrades creates a dull throb of stress. And it's only going to get worse. As Kevin Kelly notes in *The Inevitable*, the average lifespan of a new phone app is just 30 days. "Endless Newbie is the new default for everyone, no matter your age or experience," Kelly writes.

It's bad enough to imagine trudging through life on a virtual treadmill, a circle spinning endlessly against a blank screen. The more insidious problem is that life in a technology-driven, on-demand world conditions us to think in binary terms. The technology either works or it doesn't. If it works, we take it for granted. If it doesn't, we're indignant. And we think we're entitled to an immediate fix.

But in many cases the problem isn't that the technology doesn't work—it's that the technology has reached its limit. *All of our representatives are busy. Please stay on the line. Your call is important to us....*

This overcapacitization—I'm not sure that's a real word, but it fits—takes many forms. Some, like that corporate voicemail greeting, are obvious. Others are less so. And in many cases the problem isn't the technology per se, it's the infrastructure that supports it. Like the bottleneck that forms at rush hour wherever there's a lane drop on a major artery.

Limitations on technology extend beyond physical infrastructure. There are also limitations inherent in the way our technology is linked and governed. We're subject to various economic and political and legal restrictions, to a patchwork of operating systems and schedules and regulations and logistics and bureaucracies and networks and insurance requirements and terms of use, to fluctuating currency or credit systems, and to many other forces that exist only in the abstract but have all-too-tangible impacts.

As a U.S. citizen, for example, you are theoretically as free as a bird when it comes to domestic air travel. You can fly anywhere in the country you want to, whenever you choose— provided you have an enhanced driver's license that contains a Radio Frequency Identification Chip (RFID) per the Western Hemisphere Travel Initiative (WHTI) of 2009, or some other suitable form of identification. Or your credit card doesn't get declined because you are over your spending limit. Or the airline doesn't give away your seat because they overbooked the flight. Or a traffic jam doesn't delay you on the way to the airport. Or the airline's computer check-in system doesn't crash. Or a thunderstorm in Chicago doesn't delay the flight that's supposed to pick you up in Omaha and take you to Detroit. Or because of any of the countless other impediments travelers encounter on a daily basis but can't always anticipate.

~

Overcapacitized technology has been a pervasive source of stress for so long that in many cases we don't even recognize it. My first prolonged exposure to this phenomenon happened more than 40 years ago, when I was a senior in high school. I got a job washing dishes at a popular local restaurant. I had no clue what I was getting into.

Washing dishes didn't sound difficult. I'd done it plenty of times. But washing my own family's dishes, in my own kitchen on my own time, was a benign experience. The number of dishes was manageable and the job didn't take very long. Also, there was an intimacy to it. I felt protective of certain pieces. (I was careful not to break the star-spangled bicentennial glass.)

And pitching in to clean up seemed a small price to pay for a home-cooked roast beef dinner.

Washing dishes for a full shift in a restaurant was just plain awful on every level. It provided no reward other than the money it paid—which wasn't much. There was no lingering warmth from having just shared dinner with my family. I was cooped up in a small windowless room where every surface was either tile or stainless steel, cleaning up after a bunch of strangers that I had no interaction with. And the scale of the mess was staggering.

The wait staff bused the dishes to us in rectangular gray plastic buckets. Each bucket overflowed with a jumble of institutional plates and saucers and boat dishes and cups and glasses and cutlery. The tableware was marinating in a slurry of ziti and iceberg lettuce and prime rib fat and cocktail sauce and soggy garlic bread and half-used butter pats and other scraps that had been rendered unrecognizable from sloshing in cold coffee. We had to separate the garbage and dump it into industrial-size barrels, being careful not to throw away any knives, forks, spoons or linen along with it. (Sometimes it took two people to empty the garbage barrels into the dumpster out back.)

Next, we sorted the dishes and cutlery onto the appropriate racks and placed them on a conveyor belt. The conveyor belt carried the racks through an industrial-strength automatic dishwasher, which was like a miniature version of an automatic carwash. After being blasted with detergent and high-pressure jets of scalding water, the steaming racks of clean dishes emerged at the other end of the machine a few minutes later. Another of us interchangeable links in the dishwashing chain would sort the clean dishes and put them in the appropriate racks or stainless steel holders and deliver them to stations where the wait staff and cooks could reach them quickly.

Most of the time the pace was manageable. But during the weekend rush and on holidays—New Year's Eve in particular—the dishroom became overwhelmed for hours at a stretch. The automatic dishwasher could handle only so many dishes at a time, which meant the demand for clean dishes soon outstripped the supply. Any reasonable person would understand that.

There are no reasonable people at a restaurant on New Year's Eve. The crusty old waitress with the jewel-encrusted cat glasses didn't want to hear my explanations—or my *excuses*. The only thing she wanted to hear was that I could supply her with a couple of clean parfait dishes *RIGHT NOW* so she wouldn't lose that big tip she was anticipating.

Combine the overcapacitization problem with the clamor and the steam heat and the long hours, and you had an atmosphere that was as stressful as an ER or an air traffic control tower—even though the work was far less consequential. Any job is a nightmare when you feel trapped in a place where people are constantly yelling at you about things that aren't your fault.

One night, for example, the restaurant owner came into the dishroom and asked how old I was. I told him I was 17. He turned to the dishroom supervisor and told him he had to send me home at 10:00 p.m. on school nights from then on. It turned out that the restaurant had been violating child labor laws by keeping me later than that.

That night, at precisely 10:00 p.m., the dishroom supervisor bellowed my name. When I looked up, he pointed at the clock and yelled, "Get the hell out of my sight!" His tone suggested that being 17 years old was a character defect that I could have overcome if I wasn't such a pussy. And I wasn't experienced enough with the working world to realize that he was just an asshole. On some level I wondered if his contempt was justified. Maybe it *was* my fault that the dishroom crew always seemed to be under the gun.

It never dawned on me back then that most of the stress in the dishroom was a result of the automatic dishwasher's limited capacity. (That and the fact that some of the people who worked in the dishroom, like the supervisor, were trying to live on scant wages.) The restaurant owner could have relieved much of the stress by installing a second automatic dishwasher and hiring more workers. But because the times of peak capacity were relatively brief and intermittent, that solution wouldn't have been cost-effective.

That pas de deux between business owners and the people who work for them has played out in various forms, on various scales, since the dawn of the Industrial Revolution.

~

I lasted only about four months in that dishwashing job, and that was more than four decades ago. But I still recall a breakroom conversation between two guys named Tom and Tim that summarized the tyranny of technology.

Tom: I hate this job. I can't wait to quit.
Tim: So what are you waiting for? Why don't you just quit?
Tom: I can't. I need to pay off my car.
Tim: What do you need a car for?
Tom: To get to work.

Tim, a student at the college across the road from the restaurant, went on to point out Tom's circular logic. Tom, a working-class guy from an outlying town who was obviously never going to set foot in a college classroom unless he was there to clean the floor, didn't get Tim's point. I liked both of them and found the exchange uncomfortable. Tim was kind of making fun of Tom and Tom kind of realized that. In addition, I felt sheepish because I was only working as a dishwasher to pay for a car that I didn't really need, either. I feared that I would be exposed as a dilettante in the grim world of lifelong service workers.

A hardened line cook also overheard the exchange. "I have a question for you, college boy," he said to Tim. "How would you like it if I took you out back and beat the fuck out of you?"

Tim froze. His fear was palpable. "I wouldn't like that at all, obviously," he said after a moment. "And I don't doubt that you could do it. I'm afraid of you."

I was impressed by Tim's honesty. He tried to defuse the tension with a question: "What are you so angry about?"

"I don't like the way you college kids think you're better than the rest of us," the cook said.

~

Flash-forward 40 years, and that same dynamic played out on a mass scale in the 2016 presidential election. One state, two state, red state, blue state. The coastal elites vs. the neglected heartland. Powerful currents of anger and resentment flowing both ways. And too many people who seemed eager to conform to stereotypes.

As a member of the coastal elite (geographically, at least) I get why some people in the flyover states feel disrespected. Start with the term "flyover states." Snobs who say that look down on the heartland; the only time they ever see it is when it is literally beneath them. No one likes being condescended to.

But I also understand why people in the blue states were so frustrated with red-state voters in 2016. Too often, it seemed that those voters comprised an easily influenced (and highly influential) bloc that wasn't *for* anything, it was *against* everything—or at least everything connected with the vaguely defined "elite." And way too many members of that bloc seemed aggressively uninterested in facts or context or nuance. Willful ignorance isn't something you should take pride in—whether it's a blue-stater's willful ignorance of Middle America or a red-stater's willful ignorance not only of the blue states but also of the world beyond America's borders.

Incidentally, the premise that America can be neatly divided into "red states" and "blue states" is, of course, flawed. While there's a strong geographic correlation with "liberal" or "conservative" sociopolitical outlooks, the difference in those two perspectives owes more to different states of mind—actually to different types of mind—than to particular states of residence.

A better way to categorize the rift in American society is to frame it as Concrete People vs. Abstract People. Concrete People are more comfortable in situations that involve a direct physical connection to their environment and the people around them. Abstract People are more comfortable living in their own heads. And it just so happens that many of the traditional Concrete occupations—farming, manufacturing, mining—are bunched in the heartland, while the Abstract occupations—finance, advertising, entertainment, and of course the sweeping category known as "information technology"—are clustered along the coasts. So while there's a mix of each type everywhere, the concentration skews toward Concrete People in the red states and toward Abstact People in the blue states.

The danger in making such sweeping generalizations is that we stop treating people as the complex individuals that they are. I've done it myself. For example, as a writer living in New England, I'm clearly one of the Abstract People. And I've felt my share of indignation toward the Concrete People over the current political situation. At times I've dismissed huge chunks of America by painting them with a broad brush dipped in red.

And then I remember that my interaction with red-staters as individuals doesn't square with my characterization of them as a sociopolitical group. I got a slow, on-the-ground look at the "flyover states" when I took a cross-country bicycle trip in 1986. Even then, many of the issues that would come to a head during the 2016 election were already evident. Our current cauldron of political unrest didn't heat up overnight. It's been a long, slow boil.

I didn't intend to document America's deep sociopolitical divide when I set out on my bike trip. My ambition, as I noted in the journal I kept, was more self-involved: "to learn to value food, water, and shelter; to respect darkness and weather; to appreciate distance."

"I've been deprived of deprivation," I added. "What do I know of struggle, of a world that comes without running water, electricity, and instructions?" (Apparently I had qualms about our overdependence on technology even then.)

So the trip was supposed to have been as much an exercise in self-reliance as a chance to explore more of the individual states that comprised the United States of America. I didn't have a set route or timetable. Nor did I have enough money to make it to California. All I had was the harebrained notion that I could literally work my way across America. My plan was to ride until I ran out of money, pick up an odd job for a few days to get back in the black, then ride until I ran out of money again and repeat the process.

At that time I lived on Cape Cod, where low-paying jobs like painting and landscaping and construction gofer were easy to come by. In fact, I had too many of them. I looked forward to rainy days so I could get time off to write.

It was a shock to discover that even lousy jobs were in short supply once I left New England. The first time this came up was in western Pennsylvania. "You're going the wrong way if you're looking for work," someone told me. "All the jobs are back where you came from."

He wasn't exaggerating. Here's how I described the situation as I pedaled from one desperate-looking town to the next: "Few things desolate the spirit like the inability to find work. In Ohio I thought things would be better in Indiana. In Indiana I thought things would be better in Illinois. In Illinois I thought things would be better in Missouri. In Missouri I lost hope."

But it was at that point that I actually found a job—at a resort hotel similar to a place I worked at on Cape Cod. Again, we can't always color within the lines when we talk about "red states" and "blue states."

In short: The prosperity disparity between Coastal America and Middle America isn't new. The 2016 election was partly an expression of the resentment that red states have been marinating in for decades. And it has more to do with the ascension of Abstract People at the expense of Concrete People than it does with who happens to live where.

I didn't realize it at the time, but my bike trip provided a litmus test for a person's standing in American society. I noted in my journal that among those I met along the way, the reaction was "equally divided between those who envied me and those who thought I was crazy."

In hindsight I realize that almost everyone in the first category was a Concrete Person and almost everyone in the second category was an Abstract Person. And even though I was an Abstract Person at heart, I was doing something—pedaling a bicycle across the country—that defined me as a Concrete Person in that narrow context. People perceived me accordingly.

A break-room conversation between two of my coworkers at the Missouri resort summarized the divide. (It was similar to the break-room conversation between my fellow dishwashers Tom and Tim a decade earlier, although it didn't spark the same level of hostility.) Mark and Bob were both Missouri natives. Mark was a lifelong blue-collar worker. Bob was a college kid. Mark spoke with a twang. Bob sounded like a radio announcer, which was no coincidence: He worked as a DJ on weekends and hoped to have a career in a major market someday.

One day at lunch, after peppering me with questions about my trip, Mark sat back and shook his head, grinning. "Just *think* about what it is he's doing," he said to Bob. "Just *think* about it!"

"I am thinking about it," Bob said. "And I think it would suck."

As with Tim and Tom, this was the primary difference between Bob and Mark: Bob saw any kind of physical labor as atavistic drudgery that he would abandon at the first opportunity to join the more abstract and sophisticated blue-state world. Mark was just the opposite; it was important to him that his work remain grounded in a concrete realm. He just wished he could find a concrete purpose that provided a greater reward than trimming bushes at a resort in the Ozarks—which is why a cross-country bike trip sounded so appealing to him.

I think this desire to make good use of their aptitude explains why a disproportionate number of rural Concrete Americans join the military. It's not a simple matter of "finding work." It's more about finding meaning in their work.

~

The desire for meaningful work is also what made so many Concrete People susceptible to Donald Trump's vow to "Make America Great Again" in 2016. If you grew up in the shadow of a shuttered Rust Belt factory, cobbling together a collection of part-time service-industry jobs just to survive, you could easily be persuaded that you had missed out on the American heyday, when everyone pulled together to create an industrial superpower the likes of which the world had never known. You could feel that you'd been cheated out of a chance to benefit from and—just as important—contribute to that unparalleled success.

The truth is, though, that many of the blue-collar red-state workers who were in their prime during that earlier, idealized iteration of America weren't any happier than their descendants appeared to be. Dig into the history of labor relations in the automobile, steel, and mining industries—the industries that so many red-state residents seem to desperately wish would return to preeminence—and you'll find tales of strife, violence, and deep employee dissatisfaction.

Take what Sidney Fine wrote in *Sit-down: the General Motors Strike of 1936–1937.* You'd think American workers would have been happy with whatever jobs they could get during the Great Depression. Not so. As Fine wrote, "A fifty-five-year-old worker testified that the only difference he could discern between a penitentiary and the GM plant in which he worked was that the GM worker could go home at night."

Many of the worker complaints during America's industrial heyday involved the relentless pace of the assembly line. "I ain't got no kick on wages," one man said in *Sit-down,* "but I just don't like to be drove." That widespread dislike of being drove led to the GM employees joining the recently formed United Automobile Workers union. As Fine wrote, "When they struck GM and joined the UAW, [the assembly-line workers] were, in a sense, expressing the resentment of men who had become depersonalized, who were badge numbers in a great and impersonal corporation, cogs in a vast industrial machine. ... 'The world is surprised to learn,' a reporter declared shortly after the strike was over, 'that these robots are human beings after all.' "

Eighty years later, during the 2016 election, with actual robots having now taken over much of the manufacturing work, the displaced human beings appeared to pine for the good old days of depersonalization, when they had been just so many badge numbers and cogs—but at least they got paid.

~

I see clear lines connecting the red-state voters of 2016 and the dishroom workers I knew 40 years before that and the GM workers who went on strike 40 years before that. The GM workers were stressed out over technology that went a little too fast, the dishroom workers were stressed out over technology that went a little too slow, and the red-state voters were stressed out over technology that no longer needed them.

The common denominator was that the cause of their stress was technology. The relationship between the working class and automation has always been fraught.

During that whole 80-year run there had been a relatively brief time when large numbers of blue-collar workers could earn a good living in manufacturing or service industries. And that was due to a unique confluence of short-term forces. First, workers in steel, mining, automobile, and many other industries became unionized—a long, drawn-out, and at times violent process that started in the 19th century and peaked in the middle of the 20th. Workers got higher wages, better working conditions, and pension plans. ("If you get weekends off or overtime pay, thank the union members who fought for those rights," the AFL–CIO's website notes.)

At around the same time as the GM strike recounted in *Sit-down*, the Roosevelt administration initiated the New Deal programs, which not only provided jobs to many unemployed workers during the Great Depression but also introduced Social Security and other safety nets such as the FDIC. Then, shortly after that, World War II created unprecedented demand for manufacturing. (American industry produced an estimated 300,000 airplanes to support the war effort.)

When the war ended, the manufacturing sector retooled back to a civilian economy, producing the first new automobile models in more than three years. The new car culture and the thriving economy, combined with Eisenhower's burgeoning Interstate Highway System, enabled the modern suburban lifestyle that the middle class began to consider a birthright.

The long decline started in the 1970s. Gas prices climbed and the quality of American cars dropped relative to those built overseas. (The Vega, the Pinto, and the AMC Pacer became the *Nina*, the *Pinta*, and the *Santa Maria* of this voyage into mediocrity.) But there wasn't much that U.S. automakers could do to combat the stiffer competition. Poorer cars were the price of richer workers. As Megan McArdle writes in *The Atlantic*, "For decades, U.S. auto companies have been struggling with the lavish deals on wages and benefits that they made with the United Auto Workers back in the days before foreign competition broke down the Big Three's cozy oligopoly. David Cole of the Center for Automotive Research estimates that this left GM with a 'cost penalty' of more than $1,000 per car between its production costs and the competition's." That $1,000 took the form of shoddier materials and poorly cut corners.

The major reckoning finally came when the Big Three went bankrupt in 2009, and the City of Detroit followed suit four years later. Although the federal government's too-big-to-fail largesse kept the auto industry afloat, autoworkers have had to accept a new reality. Workforces have been reduced and retirees no longer enjoy a bounteous double dip of fat pension checks and Social Security benefits. Detroit, meanwhile, has become a case study in civic corruption, greed, and mismanagement. All within memory of a time "in the 1950s [when] the Detroit area had the highest median income, and highest rate of home ownership, of any major U.S. city," notes Steve Schifferes of the BBC.

Widen the lens to include the complete portrait of America's working class dating back to the start of the Industrial Revolution, and you see what an anomaly those prosperous Eisenhower years were. During the textile industry's boom years in the late 19th and early 20th centuries, workers had to fight to limit the workday to *only* 10 hours—and that was for a six-day workweek. Many textile workers subsisted in factory-owned tenements. Kids often worked alongside their parents in dangerous hellholes. In *Child Labor in America: a History*, Chaim M. Rosenberg describes conditions at the Granite Textile Mill in Fall River, Massachusetts, where 19 people died in a fire in 1874: "The children worked 12-hour days with an hour off for dinner. Nine of the 19 people killed in the fire were children 12 years of age or younger."

In short, the sentiment that helped propel Donald Trump to the White House in 2016—that the key to making America great again was to return to that sliver of time when the stars briefly aligned to create a booming manufacturing sector that enabled the lower middle class to lead prosperous, comfortable lives—was as misguided as a desire to return to the good ol' days of the California gold rush.

In fairness, however, I think the pro-Trump sentiment was a misapprehension of a much more fundamental sense of displacement. And it had nothing to do with a manufacturing economy vs. a knowledge economy or red states vs. blue states or Donald Trump vs. Hillary Clinton. In fact, it had nothing to do with America at all. I think it was actually a deep-seated urge to go back to a time that predated the United States. And it is an urge that human evolution has yet to expunge.

The skills that are often of little to no use to the working class today are the same skills that would have enabled hunter-gatherers to not only survive but also become the alpha members of the tribe. Strength. Endurance. Dexterity. Aggressiveness. The innate ability to build shelter or skin an animal. To hunt and fish. To band together to defeat a common enemy. (Incidentally, the label "unskilled," applied to anyone who lacks a college education, ranks right up there with "flyover states" for elitist condescension.)

Those skills would still have been valuable when human beings evolved to a more agrarian society, right up through the period depicted in those DIRECTV Settlers commercials. But they are not a good fit for a highly structured, highly technical, more abstract world.

Few people realize how fundamentally the Industrial Revolution has changed the way human beings experience life. It has even altered how we perceive time. Two-hundred years ago, if you asked what time it was, you would have received considerably different answers at the same moment in Boston, Buffalo, Philadelphia, Richmond, and Cincinnati—cities that are now all in the Eastern Time Zone. Back then local time really was local, with a central clock (set by observing the sun) that people synchronized their watches to. Early Americans didn't tell time so much as time told them. If you were a farmer—and 80% of Americans were in those days—the amount of daylight dictated how you went about your business.

Today mass confusion would result if every town in America went by a different time. But in the early 19th century, local time suited everyone's purpose because that's the only time that mattered. It took days to travel on horseback between major cities, so minor time differences were immaterial. And the simultaneity that telecommunications technology enabled between distant cities didn't exist yet. Face-to-face conversation was the lone form of real-time communication.

Let that sink in for a second. If you had a question that needed an immediate answer, you could get it only if the person who knew the answer was within earshot. If that person was farther away than you could shout, your answer had to wait. That's the way life was 200 years ago—and for all of history before that.

~

It was industry, not government, that introduced standardized time. When the first time zones were instituted in 1883, railroads were the driving force. Once express trains started traveling a mile a minute, that jumble of local times played havoc with schedules. Thirty-five years after that, in 1918, the federal government made standard time zones official. Hard to imagine a better affirmation of Neil Postman's contention that technology is ideology.

The impact of industrialization on modern Americans' perception of time extends beyond the division into Eastern,

Central, Mountain, and Pacific zones. Industry has also largely determined how we slice and dice our days—what time *feels* like. The eight-hour workday (and its associated rush hours). The five-day week. The weekend. The blocks of time that give our lives their rhythm. All are modern inventions. And all are arbitrary. Really, the only organic units of time are the day (the time it takes our planet to complete one full revolution) and the year (the time it takes our planet to complete one full trip around the sun). And even those units aren't absolute; they're specific to Earth. Venus orbits the sun in just 223 Earth days. But a single *day* on Venus equals 243 Earth days. In other words, a day on Venus lasts longer than a year on Venus. (Venusians must really struggle with that whole spring forward/fall back thing.)

Aside from the day and the year, every unit of time—the second, the minute, the hour, the week, the month—is an abstraction. Just don't try selling that idea to your boss when you're supposed to be at work at 8:00 and you roll in at 8:22.

A society as highly technical and interconnected as ours couldn't function without an extraordinary level of organization and strict adherence to a universal definition of time. But those requirements are also among the leading contributors to stress. People just aren't that organized or adherent by nature. (The few who are are destined to be endlessly frustrated with the rest of us who aren't.) Modern Americans don't like to be drove any more than that GM worker did back in 1937.

And really, I think that's what many Trump voters were expressing in 2016. Their votes amounted to a *Fuck you!* to an increasingly automated, complicated, regulated, abstract America. And while I understand the urge to cast a *Fuck you!* vote and even admire it on some level, I can't give Trump voters a pass. Handing such an obviously impulsive and unstable person the nuclear launch codes was an unforgivably reckless thing to do. As I'll discuss in Chapter Four, nuclear weapons are the greatest threat that man's artificial evolution has wrought. The No. 1 issue in every presidential election until the end of time should be: Can we trust that this person will do everything in his or her power to prevent World War III? And I don't see how, in good conscience, anybody could observe Donald Trump for more than two minutes and answer yes to that question.

Of course, Trump supporters didn't actually come right out and say that their votes amounted to a *Fuck you!* I think that's because many of them didn't fully understand what they were doing. They just knew they didn't like what was happening and wanted to return to some vaguely imagined better time— even though being well paid to make car parts or smelt steel or mine coal was always just a faint approximation of the kind of life that many members of the working class really wanted. They were pining for a time that predated their memories but had left its impulses encoded in their DNA. A time when being hardwired to be independent and instinctive and reactive was an asset. When the ability to think on your feet and live by your wits and adapt to changing circumstances in a hostile world was necessary for survival.

On an assembly line—where the greatest threat is that some douche-bag foreman might dress you down in front of your coworkers for violating an asinine "policy" that anyone with a lick of common sense can see is stupid—those traits could very well get you fired.

Basically, the subliminal message that tens of millions of red-state residents have received in recent decades is this: *Attention. We've reengineered the world so that people with your type of brain no longer belong. Thank you and good luck with the rest of your life.*

~

J.D. Vance, who grew up in a poor and dysfunctional family in Middletown, Ohio before winding up at Yale Law School, wrestles with some of these same themes in *Hillbilly Elegy*. Reading the story of his escape from a life of poverty, desperation, and hopelessness is like reading an account of an escape from an occupying army or a POW camp. Except that the oppressor in Vance's case was invisible. How can you resist an enemy that you can't see? Many people in such inhospitable places end up either submitting to the invisible enemy or firing shots in the dark—and some of those shots hit innocent victims.

The most illuminating episode in *Hillbilly Elegy* is an anecdote about a summer Vance spent working in a warehouse at a tile-distribution business. The job paid $13 an hour (good money in his hometown at that time) and also offered health insurance. But the owner couldn't keep the job filled. A 19-year-old kid and his pregnant girlfriend both got jobs at the warehouse but ended up getting fired in short order because "both of them were terrible workers," Vance writes.

Neither the boyfriend nor the pregnant girlfriend ever put in a full week. The guy was always late and spent a couple of hours per shift in the bathroom. (Vance doesn't say what the kid was doing in the bathroom, but my guess is that it involved a smartphone.) It was obvious to Vance that the kid was going to get fired. Still, when it happened, the kid was dumbstruck. "He thought something had been done *to him*," Vance writes.

Vance expressed shock at the kid's shock. "There is a lack of agency here," he writes, "a feeling that you have little control over your life and a willingness to blame everyone but yourself."

I see that 19-year-old kid's cluelessness as entirely predictable. That kind of cognitive dissonance is an inevitable consequence of clustering large numbers of people whose genetic resumes leave them with no sense of purpose in a technologically complex world—and also leaves them susceptible to the enervating effects of that world's many forms of distraction.

(Imagine how much less productive American industry would have been if smartphones had existed in the 1950s.)

During the 2016 election, much of red-state America acted like that 19-year-old kid in the warehouse. What they said they wanted—manufacturing jobs that provided benefits and a living wage—and what they really wanted were two different things.

~

One reason for this conflict is the compartmentalization that has come with industrialization. "Work" and "life" are separate, with little to no crossover in many cases. And when there is crossover, people often feel an uncomfortable sense of worlds colliding, like when you bring your spouse to the company Christmas party. (Many people now stay connected to their jobs 24/7 through smartphones, but in most cases this is an intrusion of their work compartment into their life compartment rather than a healthy blending of the two.)

That compartmentalization didn't exist in many earlier societies. Two centuries ago, if you had asked a Comanche what he did for work, he wouldn't have understood the question. The Comanche were nomads who followed the Buffalo herds. Moving an entire tribe up and down the Southern Plains required a constant, coordinated effort in which everyone participated. Work and life and spirituality were all tightly bound together. That's not to say life wasn't stressful. The Comanche were aggressive warriors who fought many battles. They contended with frequent outbreaks of terrible diseases, including smallpox. And, no doubt, they were as susceptible to petty squabbles as any other group of humans ever was. (*In Empire of the Summer Moon*, S.C. Gwynne writes that one Comanche name translated as "She Invites Her Relatives.") But it's a safe bet that the various Comanche bands felt a much stronger connection to one another than you'll find among Americans today.

In addition to the two main worlds of "work" and "life," Americans also inhabit lots of mini-worlds, often all by ourselves. For most of a decade I commuted 45 minutes each way to work. The 6.25% of each day that I spent alone in the car was an existence unto itself. I forged one-way relationships with radio hosts and musicians. I would go on kicks where I would buy a band's whole catalog of CDs. For days at a time their songs would play on endless repeat in my head, occupying a significant space in my conscious mind and coloring my emotional outlook. But the music was a private immersion; it wasn't an experience I shared with my coworkers when I arrived at the office or with my wife when I got home.

Until relatively recently, such a private relationship with music wasn't possible. Before the advent of recording devices, people could play or sing by themselves, but they couldn't *listen* to music by themselves. (Also, no one ever heard a rendition of a song more than once—something difficult to imagine now.) That's one of the many ways in which technology has enabled us to experience in isolation activities that we once shared.

Combine compartmentalization with that insidious isolation and you end up with a crippling lack of empathy or understanding, both between individuals and among demographic groups. The red-staters see themselves as victims; the blue-staters see them as martyrs. The blue-staters see themselves as leaders in the growing knowledge economy; the red-staters see them as smug assholes. And both sides are fiercely proud of the way they are—even though their fates were largely accidents of birth. It's like being proud of having brown hair—and having contempt for anyone who doesn't.

What few people in modern America understand is that they are as much a product of *when* they were born as where.

That thought first struck me like a hammer blow years ago, when I was working as a pinsetter mechanic at a bowling alley. I was still groping to find my way as a writer. I'd tried writing fiction but was awful at it. (This, in its entirety, was one of the rejection letters I got: "I hated this story. No thanks.") I was obsessed with writing *real* stories—so much so that I would have characters misspeak and correct themselves and say "um" and "you know" a lot. It was realistic, all right—and about as interesting to read as a transcript of small talk at the post office. Finally, I had an epiphany, although that's too pretentious a term for what happened. It was less an *Aha!* moment than a *Well, duh!* moment. I decided that if I was so interested in capturing life the way it was, maybe I should try journalism.

I had sold my first few articles while working at the bowling alley. One of my coworkers was illiterate. He knew I was a writer, which made things awkward. I think he thought I looked down on him, even though I didn't, but I knew he thought I did, so I tried not to do anything to make it *look* like I looked down on him, which might have come across as patronizing, which ... well, you get the idea. Our conversations were always stilted.

The lone exception happened on a slow night when my illiterate coworker started drawing cartoons. He was relaxed and unselfconscious as he dashed off his drawings without any apparent effort. And they were good—far better than anything I could've done. I've never had any artistic talent at all. Even today if I tried to draw a human figure it would look like something a seven-year-old would do.

At that moment I realized that if my coworker and I had both been born in an earlier time, in a society that communicated using pictographs instead of written text, he could have been a writer and I would have been functionally illiterate.

I believe that many Americans—including millions who voted for Donald Trump—are illiterate in the language of the 21st century. They're not stupid—they're simply not wired correctly for the kind of technologically sophisticated society we have now. They're trying to navigate a digital world with their analog brains, and it leaves them frustrated and resentful.

Think of it this way. Imagine if musical ability, instead of digital proficiency, was the skill most vital to getting ahead in the 21st century. A small group of people, like those with perfect pitch, or those who could sit down at a piano and play a melody by ear after hearing it just once, could thrive with little effort. Another group of people, the ones who could learn to play an instrument using sheet music, or sing in a chorus, could do well with sufficient practice. But a large group of people—the tin-eared ones who simply can't carry a tune no matter how hard they try or how many lessons they take—would be destined for a lifetime of frustration and struggle. And it would have nothing to do with their intelligence or their level of education or their sense of agency. But they would be made to *feel* that it did.

A growing intolerance, exacerbated by ignorance of our cultural history, is another side effect that the relentless pace of the Industrial Revolution has wrought. And it is a problem that is made worse, paradoxically, by our greatly improved ability to document the past.

CHAPTER THREE

Good Forgetting vs. Bad Forgetting

"It is sadder to find the past again and find it inadequate to the present than it is to have it elude you and remain forever a harmonious conception of memory."
—F. Scott Fitzgerald

The 1970s were the most turbulent 10 years of my life. I was 10 when the decade started, 20 when it ended. I'm sure the emotional fluctuations along the path from grade-schooler to adolescent to young adult were just as intense back then as they are today. My memories of those years still have a sharp edge. Everything felt so *important.*

But when confronted with visual records from that time—whether personal photographs or YouTube clips of TV shows or commercials or major league baseball games—I have a different reaction.

I think the 1970s look dumb.

Overabundant hair. Underabundant taste. Striped bell-bottoms. Avocado green and harvest gold appliances. Fake wood paneling. TVs that weighed as much as refrigerators. Cars designed with no regard for aerodynamics.

These garish superficial cues are so distracting that they diminish the emotional power of my memories. "The 1970s" looks like campy performance art.

If that's the reaction from someone who lived through that time, I can only imagine how absurd the 1970s must look to anyone born since 1980. And I can hardly blame them for not giving the 1970s much serious thought. It doesn't look like a serious decade.

This is another example of the insidious way that technology erodes our culture. And in this case, the negative impact is counterintuitive. You might think the ability to faithfully and thoroughly document the past would deepen our appreciation of it. In general, I think the opposite is true. Further, I think the ever-increasing pace of technological progress exacerbates the negative effect. New gets old fast. And we can't avoid the evidence.

~

The first photograph was taken less than 200 years ago. Until then, no one had ever seen what a younger version of themselves or an earlier version of their world looked like. (Painting and drawing have been around for millennia, but even the most accurate artist's rendering lacks the flat realism of film or video. And many paintings from antiquity didn't even aspire to realism; they had an exaggerated, hallucinatory quality that's made them a staple of *Ancient Aliens*-type shows.) Once photography and video reached the masses in the 20th century, people's perception of the past irrevocably changed. Thanks to the camera, we see images of the past as they really were, not as we remember them to have been or as they were interpreted by an artist or storyteller. That development has dealt mythology a crippling blow.

This has major implications. Mythology has been a foundational component of every culture. It's the stories a society tells itself to validate its values. And in every culture, the underpinning mythology wasn't what we would consider today to be objective truth—or anything even close to it. "All over the world," Barbara C. Sproul writes in *Primal Myths: Creation Myths around the World*, "in the Babylonians' Enuma Elish and in the earliest creed of the Celts, in the books of Job and Psalms from the Old Testament, in the myths of the Hottentots of Africa and those of the Mandan and of the Huron Indians of North America, valiant defenders of the principles of being and order do fierce battle with the forces of not-being and chaos and finally subdue them so that order and life can be established."

This struggle has been expressed in many forms, as the chapter titles of Sproul's book suggest. (They sound like they could be songs on the next Radiohead album.) *The Worm and the Toothache. Bumba Vomits the World. An Explanation of the Diagram of the Great Ultimate. A Lama Came Down from Heaven. In the Beginning the World Was Slush.* (If the Arctic keeps melting at its current rate, the world might end as slush too.)

We smirk at the very idea of mythology today. The root word, *myth*, has taken on a negative connotation not far removed from *bullshit*. No one who expects to be taken seriously in 21st century America would profess to believe in the literal truth of any creation myth, including the Book of Genesis.

But it isn't just our religious mythology that's crumbling—it's also our secular mythology. Name any great figure from early American history, and someone will be eager to tell you what an awful person he was. (And yes, most great figures from early American history were men—white men—which someone else will be eager to tell you is a big part of why America never fully lived up to its founders' lofty ideals.)

George Washington? Marvin Kitman published an entire book about how, after offering to serve without a salary as commander in chief of the Continental Army, the Father of our Country still managed to make a bundle by submitting the Mother of All Expense Accounts. (Apparently, accounting trickery among the wealthy and privileged is nothing new.) Thomas Jefferson? He owned slaves at the time he wrote the words "All men are created equal." Abraham Lincoln? His administration enforced the Long Walk of the Navajo, a 450-mile death march from what is now Arizona to the Bosque Redondo reservation in New Mexico. Christopher Columbus? He didn't discover a new world—just new ways to exploit an old one.

(In researching that last paragraph, I Googled the phrase *Was Christopher Columbus an asshole?* I got 236,000 results, including videos titled "Christopher Columbus Was a Murderous Moron" and "Why Christopher Columbus was History's Biggest Dick.")

The development of photography paralleled the development of modern American journalism (even though the two were unrelated at first and newspapers didn't start publishing photos on a regular basis until well into the 20th century). The ability to faithfully document events and capture true visual images inevitably led to a more literal, less fanciful interpretation of history. Photographic evidence also made it easier to call bullshit on previous generations.

Once upon a time, a society might have embraced tales of great deeds from days of yore. And why not? If all you had to go by was your imagination, and your society hadn't changed much in the past century, you would picture past events unfolding in a setting that looked and felt like the present, involving people who looked and acted and dressed like you. Then if you were told that those people who looked and acted and dressed like you had to overcome more severe hardships—deeper snows, greater floods, more vicious enemies—how would you feel? Inadequate, probably. Or maybe aspirational.

I hope I can be half the man my great-grandfather was when he helped fight off the Insufferable Douche Bag Tribe....

Granted, teenagers probably acted like teenagers in every culture throughout history. (Well, until recently, that is. Research suggests that, since 2012, smartphone addiction might be making teenagers more docile—and more depressed.) Bucking authority, testing limits, and bonding with your peers to the exclusion of other generations are biological imperatives at that age. (Also, it's hard to be agreeable when you're on a hormone bender that lasts several years.) Challenging your elders is a rite of passage. Your elders push back, and as you mature you realize that you could benefit from the wisdom of their experience. A century-old saying frequently (though mistakenly) attributed to Mark Twain summarizes this dynamic:

When I was a boy of fourteen, my father was so ignorant I could hardly stand to have the old man around. But when I got to be twenty-one, I was astonished at how much he had learned in seven years.

Now, however, change happens too fast, in ways that are exhaustively documented, for this dynamic to play out the way it once did. I don't think each succeeding generation ever fully gets over its adolescent sense of entitlement and superiority. And it's hard to fault young people for that. There's plenty of superficial justification for feeling as they do. Most eighteen-year-olds *are* better with new technology, and more accepting of fresh ideas, than their parents and grandparents are. Advertisers covet them. The culture covets them. And if you try to tell them that music or movies or sports were better 25 years ago, all they have to do to discredit your opinion is look at YouTube to see how unsophisticated everything looked back in the day. (And—again—I think we subconsciously equate unsophisticated technology with unsophisticated people.)

I remember feeling that way. When we were teenagers, my brothers and I watched the original *King Kong* movie (made in

1933) with my dad, who remembered it as a special-effects tour de force. He expected us to react the way he'd reacted when he was a kid—with awe and delight. Instead, we just laughed at how corny it looked. We didn't suspend our disbelief for a second. And we probably lost a little respect for my dad's judgment. How could he have been taken in by something so phony and primitive?

Cameras have existed long enough now that even older people respond with condescension or dismissiveness to images from the time long before they were born. Viewed on film, the world around the turn of the last century is too quaint to feel real. I love reading about baseball history, but when I look at ballpark photos from the early 20th century, I don't feel an emotional connection. I can't imagine what all those men dressed in suits and identical straw hats did all day. They look less like real people than extras in a movie about turn-of-the-century baseball.

~

Upshot: The tethers that create a meaningful connection to the past get shorter, and the anchor points get weaker, with each generation. Mine are probably a lot longer and stronger than most Americans'. I claim no special credit for this; it's just another accident of birth.

Although it ended 14 years before I was born, World War II was the creation myth for the modern world as I understand it. The war was the last time America's reality conformed unambiguously to the creation myth's basic architecture—Sproul's valiant defenders of the principles of being and order battling with the forces of not-being and chaos and finally subduing them so that order and life can be established (or reestablished, in this case). In World War II, we were clearly the good guys, and the good guys won. Everything that has followed, followed from that.

Growing up, I indulged my fascination with World War II through popular culture. (*Snow Treasure* was one of the first books I remember reading on my own.) I committed significant dates to memory without trying: Pearl Harbor Day, D-Day, Hiroshima and Nagasaki. My knowledge of World War II also gave me a basic understanding of the world I grew up in—the way the abrupt end to the shotgun wedding between us and the Soviets led to the nuclear arms race and the Cold War, the way our efforts to rebuild Germany and Japan helped shape the modern world economy, and so on.

In addition to cultivating a solid historical perspective on World War II, I also developed a vicarious emotional connection to it through my parents. They grew up in Glasgow during the war and told a lot of stories about it. (And because they had no photos from those days, I relied solely on my imagination when picturing their circumstances.)

My dad once told me that he remembered fearing for his life as he huddled in an underground shelter, listening to the drone of Luftwaffe planes and feeling the concussions of the German bombs. There were numerous industrial targets along the River Clyde, but many bombs also fell in random locations elsewhere, sometimes wiping out entire neighborhoods. My dad said his legs were shaking so violently during one raid that his knees were literally knocking.

Like my dad, my mom lived with rationing and shortages that went on for years. She and her entire family (six in all) lived in a two-room tenement. One summer she was sent to live in the country with complete strangers. (Better to keep the kids out of harm's way.) After the war she couldn't warm up to Germany as an ally.

And she carried emotional scars from the bombing, too. When I was a kid, our hometown occasionally tested the air-raid sirens during those ludicrous Cold War "duck and cover" civil defense drills. The sound of those sirens induced post-traumatic stress in my mom. I could see it in her face. Once we were walking into town when the sirens went off and she froze.

I've gone through life assuming that most Americans feel the same deep connection to World War II that I do. The older I get, however, the more I realize that many younger people have only a dim knowledge of the war, and no emotional connection to it at all.

My work on an article for *Down East* magazine drove this home. It was about a man named Peter Noddin, who had gone to extraordinary lengths to document the location of every military plane crash in Maine. Your reaction might be: *Every military plane crash in Maine? How many could there be?* The answer astounded me: 805, more than half of which happened during World War II, resulting in 143 deaths.

The research and reporting on that story drew me in unlike any other magazine article I've ever written. Maine was both a training ground for aircrews and the last stop on the U.S. mainland for many of the warplanes ferried to Europe by way of the Arctic Circle during World War II. Because of the urgent need to supply air support for the Allies, there wasn't enough time to ensure quality control on the planes (which were comparatively primitive to begin with), to thoroughly train the crews, or to wait for perfect flying weather.

As a result, fatal crashes were appallingly common. Peter Noddin's database led me to one harrowing story after another. The Royal Canadian Air Force plane that was diverted from Halifax to Montreal in November 1940 because of bad weather but ran out of fuel and crashed along the Maine–Quebec border. When it was time to bail out, the six-man crew discovered that they had only five parachutes. Two men shared a chute; they drowned when they crashed through thin ice on a remote lake.

The A-26 Invader that aborted an attempted landing in Portland in July 1944 and crashed in a trailer camp housing shipyard workers. The crash and resulting fire killed the two-man crew and 17 civilians. (The spot where the plane hit is now a swimming pool in a condo complex.)

The A-26B that crashed just outside Bangor in September 1944 because a jury-rigged tank failed and starved the engines of fuel. The same problem had happened to the same crew on the same plane the day before, but that time the pilot had managed to restart the engines. "Switched over to the bomb-bay tank after we got up to 3,000 feet but something was wrong with the piping system," the navigator wrote in a letter that arrived home after he died. "Both engines quit, and so did our hearts, almost."

And then there was the pair of B-25C bombers out of Presque Isle that crashed just 15 miles apart on the morning of September 22, 1942. Although visibility at the airfield was sufficient for Europe-bound planes to take off, there was rain and heavy cloud cover throughout the surrounding area. The first plane went down not long after its 7:25 a.m. takeoff, killing all seven crewmembers. Army investigators later concluded that the pilot had become disoriented in the thick soup and inadvertently put the plane into a steep dive that he could not recover from.

The war machine paused—briefly. Desperate to get as many planes as possible to Europe before winter weather suspended operations, the Army resumed flights out of Presque Isle just two hours after the crash. Almost immediately, another plane went down near a potato farm, in all likelihood for the same reason as the first, killing another seven crewmen.

The wreckage of the first B-25C, in a cedar swamp in the North Woods, is still there. Peter Noddin took me to the spot. There's no formal memorial or marked path. You need a GPS to find it. As I wrote in the article, "Picking through the brush takes focus. Then, the first glimpse of the wreckage changes the mood. The excitement of discovery—*there it is!*—is tempered by a sense of intrusion."

The wreckage was so well preserved after 75 years that it was spooky. Peter, who had been to the site before and had documented every piece of debris, pointed out the impact crater, now filled with water the color of iced tea. He said a bear had once used part of the fuselage as a den.

He pointed out a flat piece of rusted metal that had stumped him during an earlier visit because it didn't look like any airplane part that he was familiar with. Then he figured out what it was: a toolbox that had been flattened on impact. Right next to it was a metal bucket that, though thoroughly rusted, was intact. Peter speculated that the Army recovery team had left the bucket behind when they retrieved the crewmen's bodies.

As I wrote in the article, "The bucket and the toolbox—an artifact of the living alongside an artifact of the dead—transformed the B-25C crash from a distant historical abstraction into something immediate and tangible."

I had always been in awe of the courage of those who had fought against the Axis powers. (I remember lying awake in bed after watching a World War II movie as a kid, wondering if I would be able to display the same level of selflessness if called upon.) Peter's research made me realize that those displays of courage extended well beyond the battlefields of Europe and Asia. Standing at that crash site in the remote Maine woods brought World War II home to me in a new way. I tried to imagine what those seven crewmen must have felt when they took off that September morning into the enveloping cloud cover. Their proposed route from Presque Isle to Prestwick, Scotland would have required dicey stops at bases in Canada, Greenland, and Iceland. (It was similar to the path of the first jet flight I'd made as a kid, but that flight was nonstop in a pressurized cabin at 35,000 feet.) And if they survived that trip and made it to Britain, *then* they could start flying bombing missions over Germany. The crewmen had to know they were taking tremendous chances with their lives—but they took them anyway because the stakes demanded it. I've never been in a situation even remotely similar.

~

I notice different reactions along generational lines when I tell people about visiting that crash site. Those my age or a little older want to know more. But when I mention it to people a generation or so younger than I am, the reaction is always along the lines of, "Really? Huh." And that's it. No curiosity or follow-up questions. No connection.

That's not a criticism. It's just an observation about how quickly the world moves on in an age of rapid technological (and cultural) change.

I do it too. Not long after I wrote that article, I read Nathaniel Philbrick's book *Bunker Hill*. I wanted to know more about how America began. So many of the decisive moments of that seminal time had happened not far from my home in New Hampshire, and I knew little about them. The book was filled with the names of places I've been to—not just Boston, but also Andover, Cambridge, Danvers, Dedham, Framingham, Marshfield, Plymouth, Portland, Portsmouth, and many others. You'd think I would have felt the same sort of connection I'd felt when researching World War II crash sites in Maine—a new appreciation for the violence of war that had erupted right in my neighborhood.

But I didn't. The world Philbrick described was simply too remote and too primitive for me to imagine. Back then, even well off people subsisted in conditions that we would associate today with privation, such as the lack of indoor plumbing. Communication with England took months. Officers in the field communicated by dispatching messengers on horseback, who used language like this: "This hue and cry occasioned a sallying forth of the people from the houses ... others upon the road joined in the cry—all endeavoring to stop me."

As a collection of facts, the book interested me—in a "Really? Huh" sort of way. But I didn't *feel* it. I couldn't envision the horror of grapeshot raining down on Charlestown the way I could envision the horror of a plane plummeting from the rainy skies into the Maine woods.

~

When journalists grope for perspective on current events, a common device is to imagine what people will say about those events in 100 years. That's easy for me. I don't think people a century from now will say anything about this time. Provided humans are still around at that point, the overwhelming majority of them likely will know very little about the early 21st century, much less care about it. Those tethers to the past will keep getting shorter and shorter.

There's evidence that technology, online technology in particular, is eroding not only our attention spans but also our ability to remember things. I've long suspected that it's happening to me. But it's hard to say for sure because forgetfulness is also a function of age and I'm 60 years old.

Some people blithely suggest that even if the internet is interfering with our memories, it's no big deal. Why waste those encephalic gigabytes storing facts and figures that you can Google anytime you need them?

The problem with relying on externalized memory, as it's sometimes called, is that you have to remember the need to remember. And you might not always do that.

Nicholas Carr explores the relationship between memory and technology at length in *The Shallows: What the Internet Is Doing to our Brains.* He reaches some disturbing conclusions. The process by which our brains transfer short-term memories into long-term memories is extremely complex. It's also easily interfered with. Instead of augmenting memory, Carr concludes, the internet "places *more pressure* on our working memory, not only diverting resources from our higher reasoning faculties but obstructing the consolidation of long-term memories and the development of schemas. ... The web is a technology of forgetfulness."

Among other things, Carr writes, "The influx of competing messages that we receive whenever we go online not only overloads our working memory; it makes it much harder for our frontal lobes to concentrate our attention on any one thing. The process of memory consolidation can't even get started."

Carr sounds a warning: "To remain vital, culture must be renewed in the minds of the members of every generation. Outsource memory, and culture withers."

~

Photography and the internet. Two forms of outsourced memory, each with a specific corrosive effect on our ability to maintain solid emotional connections to the past and a shared sense of cultural heritage. Combined, they've spun off another soul-deadening scourge, exemplified by Snapchat.

With Snapchat, photography has come full circle. Before photography, there was no way to capture true visual images for future reference. Photography gave us an unprecedented ability to document events in our lives. And for many years, that was photography's primary purpose. We took pictures while on vacation so we could have them developed when we got home and keep them in albums (or, more often than not, stuffed in the envelope from the drugstore, unsorted). Then we could look at them and remember the warmth of Key West after we returned to the deep freeze of midwinter. (At least until a few years passed and our reaction leaned more toward "Look at that awful mustache I had back then!")

Things changed when photography evolved from film to digital. Because film was more cumbersome and more expensive, we tended to be more judicious about what we shot. Digital photography eliminated restraint. We could shoot as many pictures as we wanted at no additional cost. And because we could review them instantly, we could delete the duds and save the rest for ... what, exactly?

Older people like me, who thought of photographs as tangible keepsakes, struggled to adapt to the new reality of digital photography. Where do you store all those images?

The answer, in more and more cases, is: You don't store them anywhere. Photographs aren't for keeping—they're for sharing. Snapchat took this idea to an extreme by making it so you *couldn't* keep photos.

If you accept my earlier premise that photography's unsparing fidelity has lessened our appreciation of the past, then you might think I would welcome this shift toward fleeting, disposable images. But I don't. The problem with apps like Snapchat is that they celebrate the present at the expense of *being* present. You're sharing the moment with somebody who is someplace else. And as a recent study suggests, this has a negative effect. Actually the study does more than suggest; the title baldly states the problem: "How the Intention to Share Can Undermine Enjoyment: Photo-Taking Goals and the Evaluation of Experiences."

The study found that when people take photos for the purpose of sharing them on social media, their perspective tends to shift from the first person to the third person— from the perspective of a participant to the perspective of a detached observer. "When people are in more of a third-person perspective, they'll have less intense emotions when they relive the experience," Alixandra Barasch, a cognitive scientist at NYU and one of the study's authors, told VOX, "whereas if I stay in the first-person perspective, I feel the genuine emotions that I felt during the exchange."

Like so many other changes wrought by technology, sharing selfies through social media is having a profound impact on society. The change has also happened quickly—and yet almost imperceptibly. "Fifteen years ago," Taylor University professor Zack Carter wrote in a 2017 *Psychology Today* article, "if you were to take your Nikon CoolPix camera and begin taking photographs of yourself, sending them to your friends and family every day, you'd be labeled some sort of a lunatic."

~

Ubiquitous cameras have led to other forms of lunacy, involving video as well as snapshots. Besides creating a nation of selfie-absorbed narcissists, smartphones have spawned a surveillance state. But it isn't Big Brother who's watching, as George Orwell predicted. It's millions of Little Brothers.

Sometimes these citizen stringers armed with cellphone cameras do some good, as when they capture acts of police brutality. But mostly the effects have been negative. Among other things, all those cameras have chased public figures, including politicians, underground—to the detriment of our democracy. Almost nothing of consequence happens where anyone can see it anymore. The big decisions get made behind closed doors. The stuff that happens in public is pure theater, staged just for the cameras.

Meanwhile, many spontaneous events that now pass for "news" are of no importance. Watch your local station at 6:00 and 11:00 and much of what you'll see is a digest of the day's most popular social media posts. I live in New England— what do I care about a sinkhole in Florida or an inept convenience store burglar in Colorado or a cute animal story from Minnesota?

Producers arrange programming based on the best available video, not the relative significance of events. If a drunken celebrity mouths off to a cop, the bodycam video of the exchange will be treated like a major breaking story. Meanwhile, something like passage of the Gramm–Leach–Billey Act, which set the stage for the financial meltdown of 2008, goes virtually unreported.

The abundance of inexpensive, sophisticated video technology has had other negative consequences as well. Employee surveillance, for instance. (Almost half the companies in a recent survey say they now use cameras in the workplace.) Yes, video cameras can reduce employee theft, and in the case of convenience store employees, can improve worker safety.

On the other hand, there's evidence that cameras in the workplace lead to increased stress and higher rates of employee turnover. And no wonder, when groups like the Society for Human Resource Management encourage companies to adopt policies that use language like this: "Employees should not have any expectation of privacy in work-related areas." (Canadians seem to care more about this issue than Americans do. In Ontario, a woman filed suit after finding that her employer had installed a security camera in her office without telling her. The Ontario Superior Court acknowledged that the woman had no inherent right to privacy at work, and yet awarded her damages anyway because the secret camera had created a "poisoned workplace." Good luck winning that argument in the U.S. courts.)

In other industries, the negative effects of using video technology are more insidious. Take high-definition replay technology in sports. Televised games have become increasingly unwatchable because replay reviews drag on far too long and disrupt the continuity. And in many cases the replay reviews don't result in clear-cut decisions—they simply push the arguments over interpretations of rules to a more abstract level. If you had asked James Naismith, the inventor of basketball, to explain the difference between a Flagrant 1 Foul and a Flagrant 2 Foul, he wouldn't have had the faintest idea what you were talking about.

But once you start using video technology, it's hard to stop. The same is true of all technology—even when the consequences of using that technology pose a clear existential threat.

CHAPTER FOUR

Nuclear Weapons Don't Kill People, People Kill People

"Since I was 14, the overriding objective of my life has been to prevent the occurrence of nuclear war."
—Daniel Ellsberg, Hiroshima Day, 2009

On August 6, 1914, during the first days of the First World War, a German zeppelin conducted the first-ever airstrike against a civilian target, the Belgian city of Liege. An estimated nine civilians died.

On August 6, 1945, in the dying days of the Second World War, the *Enola Gay* detonated the first atomic weapon, the uranium bomb Little Boy, 1,900 feet above the Japanese city of Hiroshima. An estimated 78,000 civilians died instantly.

It took just 31 years, to the day, to make that leap. Less than half a human lifetime to progress from a scheme that was childlike in its lack of sophistication—*Let's throw bombs at people from a great big balloon!*—to a weapon that picked the lock on the secrets of the universe.

When it comes to the forces that drive technological innovation, war has no equal.

~

There's nothing new about warring nations killing civilians en masse. Massacres date back millennia. So do genocides. The 20th century's two world wars had both, including the worst genocide ever recorded, the Holocaust.

But much of the horror unleashed against civilians in those two wars was of a different type than history had ever known. It was committed with neither malice nor regret but instead with cold indifference. Civilian deaths were simply a side effect of a new approach to war, fought at a remove using weapons of mass destruction (a term first used in 1937). "Collateral damage," those civilian deaths came to be called.

You can trace the rationale for a more efficient, industrial form of war to that first aerial assault of World War I. It was sparked by a chain reaction that had begun when Archduke Franz Ferdinand of Austria and his wife, Sophie, Duchess of Hohenberg, were assassinated in Sarajevo on June 28, 1914. That led Austria–Hungary to declare war on Serbia a month later. Serbia had a mutual-support treaty with Russia and France; Germany had a similar understanding with Austria–Hungary. So suddenly Germany found itself at war with both France and Russia.

Germany felt an urgent need to defeat France quickly so it could focus on the more daunting Russian front. Its strategy, called the Schlieffen plan (after German Chief of Staff Alfred von Schlieffen), was predicated on a quick strike on the north of France. This plan required sending German troops through Belgium, a neutral country. One problem: Belgium refused Germany's request for passage. So Germany invaded. Now a quick strike in France would require an even quicker strike in Belgium.

The first obstacle was the heavily fortified city of Liege. The Germans tried overwhelming the city with sheer numbers. They suffered catastrophic losses, with casualties estimated at more than 5,000. Many German troops were slaughtered by machine-gun fire. That's what happened, with horrifying regularity, when old-school tactics, like charging in large numbers across open spaces, ran up against new mass-killing technology.

If an airstrike from a zeppelin could take out entrenched defensive positions faster and help stanch those massive losses, then inflicting incidental death on a handful of civilians seemed a small price to pay. Thus began the slippery slope toward using weapons of mass destruction in order to satisfy a perceived "greater good."

Soon German commanders recalibrated their thinking regarding civilian deaths from bombing raids. They began to see them not as an unfortunate side effect but as a desirable outcome. Hit the enemy where they live, the thinking went, and you will undermine their morale. The Germans believed airstrikes could be a particularly effective technique against Great Britain, because the bombings would rupture the sense of security Britons felt living on an island protected by the Royal Navy, largest in the world.

"The measure of the success [of a bombing campaign] will lie not only in the injury which will be caused to the enemy," Admiral Alfred von Tirpitz wrote in a letter toward the end of December 1914, "but also in the significant effect it will have in diminishing the enemy's determination to prosecute the war." That, in turn, could save many, many more lives of both soldiers and civilians in the long run by shortening the war. *The greater good...*

Less than a month after von Tirpitz wrote that letter, the Germans launched the first of 52 long-range zeppelin bombing raids over England. The raids, which were conducted at night, resulted in more than 500 civilian casualties, children included. Instead of breaking the Britons' will, however, the zeppelin raids deepened both their resolve and their hatred of the enemy. ("Baby killers," they called the zeppelins.) Britain also figured out how to fight back—and the big, combustible balloons were easy targets.

Instead of conceding that its strategy was flawed, Germany upped the ante by developing a line of heavy bombers.

~

Airplanes were rickety and rudimentary when World War I began. It had been barely a decade since the Wright Brothers had first taken flight at Kitty Hawk. Even so, it was immediately apparent that aircraft would change the nature of war. "The remarkably definite way in which positions and movements of the German troops have been located by the general staffs of France and Belgium is due almost entirely to the success of aerial reconnoitering," *The New York Times* reported just two weeks into the war. The unnamed correspondent's description evoked the same technological evolution evident in the Miles Brothers film *A Trip Down Market Street:* "The advent of the aeroplane has already revolutionized strategy and tactics. ... Reconnaissance in force by cavalry has been almost superfluous."

Other than providing superior surveillance, airplanes weren't equipped to be effective offensive weapons in 1914. From the "aeronautical correspondent" at London's *Daily Telegraph:* "The opinion generally held by experts [is] that the small bombs which alone can be carried on an aeroplane are incapable of doing serious damage."

The airplane's ability to inflict damage rapidly improved. One of the first major developments was the synchronized machine gun, which inventor/entrepreneur Anton Fokker helped introduce in 1915. This enabled biplane pilots to fire from the cockpit at targets directly in their line of sight, by shooting bullets between their spinning propeller blades without hitting them. (It didn't always work. And a French pilot, Roland Garros—if the name rings a bell, it's because the tennis stadium where the French Open is played was named after him—fired an unsynchronized machine gun through his propeller blades. He put wedge-shaped pieces of armor on the blades to deflect bullets and just hoped he wouldn't shoot his own prop off. The courage of those early flying aces is difficult to fathom.)

The synchronized machine gun improved the airplane as both a defensive weapon, capable of shooting down enemy reconnaissance planes, and as an offensive weapon that could strafe ground troops.

By 1916 dedicated bombers had appeared. Germany's Gotha heavy bomber was capable of taking out strategic targets such as railroad bridges. That November Germany also launched its first fixed-wing bombing attack on London, directed at Victoria Station.

On June 13, 1917, 14 Gotha G.IV bombers struck London in a daylight raid as part of Operation Turk's Cross. This single raid resulted in higher casualty totals than all of the zeppelin attacks combined to that point: 162 dead and 432 injured. The death toll included 18 children, most aged 4–6, at Upper North Street Elementary School, which was struck by a wayward bomb apparently intended for East London's docks.

That was the first full taste of what would become standard 20th century warfare.

~

The war to end all wars didn't. And when widespread hostilities inevitably erupted again two decades later, it was apparent that whoever controlled the skies would control the world.

Under the Treaty of Versailles, the vestiges of Germany's air force were dismantled—supposedly for good—in 1920. But that was also the year that the German Workers' Party changed its name to the Nationalsozialistische Deutsche Arbeiterpartei. Nazis, for short.

Treaty, schmeaty.

By 1935 Germany had a newer, bigger, and much deadlier air force, the Luftwaffe, to support the Nazis and their leader, Adolf Hitler. When the Nazis invaded Poland on September 1, 1939, the Luftwaffe consisted of about 4,200 planes, including more than 1,500 bombers. Those numbers would soar in the next five years, peaking with an output of some 35,000 planes in 1944.

For much of the war's first year, German bombers concentrated on industrial or military targets, particularly Royal Air Force fields in Britain. Then, on August 24, 1940, the Luftwaffe struck central London. Because Hitler had forbidden a direct assault on the British capital, many World War II historians believe that attack was an accident, a result of off-course bombers misidentifying their target.

Regardless, the damage was done. There were civilian casualties, and among the structures hit was St. Giles Cripplegate, the 14th century church where Oliver Cromwell was married and John Milton buried. The church had survived the Great Fire of 1666—but not a Nazi bomb that took out the northwest wall. *The New York Times* noted it was sadly fitting "that the enemy of civilization and religion should strike to batter down an ancient church and to destroy memorials precious to literature and liberty."

Britain's new prime minister, Winston Churchill, was incensed. Although hundreds of British civilians had been killed in raids on industrial targets, this attack on a cultural center with no military value seemed like a wanton breach of decorum, even for Hitler. The next day, the RAF retaliated by bombing Berlin.

Now Hitler was incensed. He launched a tirade against the British and vowed to "eradicate their cities."

From then on, all pretense of civility vanished. In this new type of war, everyone was a participant. The Luftwaffe blitz began in September and continued almost without letup until the following May. At one point, London was bombed for 57 straight days. In the end, there were 43,000 British casualties, and a quarter-million homes were destroyed. A single attack on May 10 killed 1,436 Londoners. In Coventry, 550 people died on November 14, 1940. In Scotland, a raid on the nights of March 13 and 14, 1941, left 650 dead in Glasgow and another 528 dead in Clydebank, where more than a third of the city's 12,000 houses were destroyed.

In that raid on Glasgow, a girl who was about to turn 10 and a boy who was about to turn 11 were spared, but psychologically scarred for life. Eighteen years later they became my parents.

~

Almost every war is, to some degree, a war of attrition. The side with the most troops, equipment, food, water, and resolve will eventually win.

That's why, from the moment the Japanese launched their surprise attack at Pearl Harbor on December 7, 1941, it was almost inevitable that the Allied forces would win World War II.

Until then, the U.S. had hoped to remain neutral. The sneak attack not only brought the U.S. into the war, but also did so with the near-unanimous support of the American people.

The U.S. was still climbing out of the Great Depression. Ramping up its factories and farms to full production (and full employment) to fight a pair of hated enemies was an easy sell. And unlike our enemies (or most of our allies), our civilians were insulated from the fighting by a vast ocean on either side. In addition, the U.S. was fresh, while Japan had been fighting

China since 1937. And even though Germany had seized control of Western Europe by December 1941, it had paid a heavy price in lives and equipment, losing almost 2,000 planes (and many of its best pilots) in the Battle of Britain alone. Axis factories struggled to keep pace with the Luftwaffe's losses.

It took 3½ years, but U.S. industrial output eventually won out. America produced more planes in 1944 alone than Germany did for almost the entire war. By 1945 the Allies were conducting massive bombing campaigns almost unopposed by depleted Axis forces.

That March, a squadron of 334 B-29 Superfortresses dropped incendiary M-69 cluster bombs on a Tokyo suburb. The bombs contained tubes of napalm, a product of American ingenuity. Napalm was invented by Harvard professor Louis Fieser and produced by U.S. petroleum and chemical companies. A *Time* correspondent described M-69s as "miniature flamethrowers that hurl cheesecloth socks full of furiously flaming goo for 100 yards."

Just past midnight on March 10, 1945, all that flaming goo sparked a wind-whipped blaze that burned 15 square miles of Tokyo to the ground and killed as many as 130,000 civilians. A similar firebombing raid a month earlier had incinerated Dresden, "the Florence of the Elbe."

It was obvious by then that the Allies were dominating the war of conventional weapons. (And yes, by 1945 the definition of "conventional weapons" had expanded to include cheesecloth socks filled with burning jelly.) Less obvious was which side was ahead in the race to develop unconventional weapons capable of almost incomprehensible destruction. In June 1944 Germany had begun using V-1 "flying bombs." These were crude cruise missiles—unmanned jet aircraft built with sheet metal and plywood and powered by gasoline. Even so, they created havoc, killing 6,000 people and inflicting widespread damage. And they ramped up the level of terror because they could strike any place at any time in any weather, accompanied by a chilling sound (they were nicknamed "buzz bombs").

The V-1 had limited range, and by fall the Allies had captured all of the launch sites. But by then the Germans had unleashed the V-2, a 46-foot-long liquid-fueled rocket armed with a ton of TNT/ammonium nitrate mixture. The V-2 had a longer range, reaching targets 200 miles away in just five minutes. It soared to a height of 50 miles, then plunged to the earth at speeds reaching 3,500 mph. No airplane or antiaircraft gun could intercept it. In addition to a massive explosion, the V-2 also produced a sonic boom.

In the first V-2 attacks on September 8, 1944, Germany struck both Paris and London. And although the first V-2 attacks were wildly inaccurate, the thought of an increasingly desperate Adolf Hitler armed with unstoppable ballistic missiles was terrifying. Particularly if the Nazis developed an atomic warhead.

~

The concept of nuclear weapons had originated in Germany in 1938. From then on the Allies and the Nazis were in a race to develop the first workable atomic bomb. In hindsight, it's clear that Germany developed its ballistic missile program at the expense of its nuclear research. The U.S. did the opposite. America funneled $2 billion into the Manhattan Project, which culminated on July 16, 1945 with the first nuclear explosion, at the Trinity test site in New Mexico.

But by then Germany had surrendered. Hitler was dead and the war in Europe was over. The war in the Pacific continued, but Japan was in tatters and posed no known nuclear threat.

So why did the U.S. drop an atomic bomb on Hiroshima on August 6, 1945? The complicated answer came down to the same "greater good" logic that inspired the first zeppelin bombing raid at Liege 31 years earlier.

By that point Japanese leaders realized that there was no realistic chance of victory. So their goal became to prolong the war as long as possible, while inflicting maximum losses. The hope was to force a stalemate in which the Allies would agree to allow Japan's power structure, led by emperor Hirohito, to remain intact after the war.

This strategy produced a horrific outcome. In June and July 1944, U.S. Marines capture the island of Saipan, surrounding 4,000 Japanese troops led by Lieutenant General Yoshitugu Saito. Rather than surrender, Saito ordered his men to mount a suicide attack. (Saito killed himself before it happened.) Amid chilling cries of "Banzai!" the doomed Japanese troops charged the Marines and managed to take 1,000 Americans with them. In addition, an untold number of civilians, fueled by horror stories about savage Americans, threw themselves off cliffs rather than be taken prisoner. When Japan's leaders hailed these martyrs as heroes, an expectation was set for the rest of the war: Soldiers and civilians alike were to give their lives to uphold the honor of the emperor.

This resulted in another tragic outcome in the Battle of Okinawa in the spring of 1945. As at Saipan, the Japanese commander, Lt. Gen. Ushijima Mitsuru, committed suicide —but not before ordering all those under his command to do the same: "Every man in these fortifications will follow his superior officer's orders and fight to the end for the sake of the motherland. This is my final order. Farewell." More than 100,000 Japanese troops died. Kamikaze pilots attacked the U.S. fleet, sinking 36 ships and inflicting almost 10,000 American casualties.

Tens of thousands of civilians also died. As the Americans closed in, Japanese forces supplied the local population with grenades that they could use to kill themselves. But the grenade explosions spooked American forces, who fired on the civilians. That only reinforced the civilians' fears of the merciless Americans, and the spiral continued.

One of those civilians was 16-year-old Kinjo Shigeaki. In the book *Descent into Hell: Civilian Memories of the Battle of Okinawa*, Shigeaki recalled being knocked unconscious by a grenade blast. When he regained consciousness, he saw a local politician beating his wife and children to death with a tree branch. Shigeaki and his older brother followed that example and killed their mother, sister, and youngest brother. The two of them then faced an appalling dilemma. The book describes the scene:

Kinjo was both afraid and worried about being the last one left alive. "Who was now going to kill whom?" His mother, sister, and brother were already dead and he and his elder brother had to decide which of them would be next. As they were discussing it, two people from the same year at elementary school on Maejima stepped between them and said, "Look, if we're going to die, let's die attacking the Americans."

They thought that to go out and launch a suicidal attack against the Americans would be the best way to die. "We thought that, as the last living citizens of the Empire, we each had to take an enemy soldier with us when we died. With that, we agreed on an infiltration attack, challenging ourselves to an even more frightening death." For whatever reason, two sixth grade elementary school girls joined them. Five others aged between 12 and 19 also found themselves included in the infiltration squad, armed only with sticks.

U.S. leaders now faced their own appalling dilemma. Allowing Hirohito to remain in power (power that many Japanese believed to be divine) was unacceptable. But trying to seize control of the Japanese mainland in the face of 70 million people who were committed to killing themselves rather than surrender—and to try to take as many Americans with them as possible—could result in a protracted bloodbath. Maybe a new bomb that could vaporize an entire city in a single unholy flash was the only thing that could break the spell of Japan's fanaticism.

It was left to Harry Truman, a failed haberdasher born in rural Missouri in the 19th century, to perform the atomic bomb's thankless calculus. Truman was an accidental president who had taken office just four months earlier when FDR died. And yet fate had tapped him to make one of the most momentous decisions in the history of warfare.

To his credit, Truman didn't even try to portray the instantaneous death of 78,000 people as a small price to pay to achieve "a greater good." But he still declared it an *acceptable* price. "We won the race of discovery against the Germans," Truman said in a radio address. "We have used [the atomic bomb] in order to shorten the agony of war, in order to save the lives of thousands and thousands of young Americans."

He later estimated that the bombing of Hiroshima (and subsequent bombing of Nagasaki) saved 500,000 American lives—more than the total number of Americans who actually died in World War II.

Still, Truman was under no illusion that using weapons of mass destruction to shorten World War II would save humanity ever after. After touring the bombed-out remnants of Berlin, he'd made this entry in his diary: "Never did I see a more sorrowful sight, nor witness retribution to the nth degree. It is the Golden Rule in reverse—and it is not an uplifting sight. What a pity that the human animal is not able to put his moral thinking into practice! ... I hope for some sort of peace, but I fear that machines are ahead of morals by some centuries and when morals catch up perhaps there'll be no reason for any of it."

Truman wrote those words on July 16, 1945, the date of the first nuclear explosion at the Trinity test site.

~

Truman's fear was well founded. Not only did nuclear capability not ensure a lasting peace, but it also led humanity directly into the most perilous quarter-century it has ever known. Part of the reason the world was in such great danger was that military leaders had become desensitized to massive loss of life and property. Little surprise in the wake of a war that had killed 70 million people and devastated many of the world's great cities.

To appreciate just how fraught this period was, you have to understand the context. First, the U.S. had developed the atomic bomb not as a deterrent but as an offensive weapon. Yes, the hope was that *once it was used* it could serve as a deterrent. But its true purpose was to end the war in the Pacific in the most expedient way possible.

And it worked. So why wouldn't nuclear bombs work again to achieve an important military objective?

To generals in the field, war had increasingly become a matter of technological efficiency. The atomic bomb was the most efficient weapon ever devised. One plane armed with one bomb could achieve the same result as a hundred planes armed with several thousand bombs while exposing far fewer pilots to risk.

It's easy to see where this thinking could have led if left unchecked. That's why, after the bombings of Hiroshima and Nagasaki, Truman declared, as commander in chief, that no further use of atomic bombs would be allowed in Japan without his order.

Still, there was no official protocol in place. And immediately after World War II came a tense period when U.S. leadership was basically making up a policy on the fly for dealing with its new doomsday technology. The first real test came in Korea, which *Air & Space/Smithsonian* magazine called "nuclear kindergarten."

Not even five full years had passed since the end of World War II (in September 1945) and the start of the Korean conflict (in June 1950). But the world had changed dramatically. The U.S. and the Soviet Union, allies of convenience in World War II, were now waging a cold war to determine whose economic and social philosophy—capitalism or communism?—would dominate the world.

Leaders on both sides wanted to avoid an all-out military conflict that would embroil the entire planet yet again. So what developed instead was a series of highly centralized battles by proxy (including Vietnam and Cuba), starting when communist North Korea crossed the 38th parallel and invaded capitalist South Korea. The Soviets and the Chinese, the world's two leading communist powers, supported North Korea; a United Nations coalition led by the United States supported South Korea. But neither side actually declared war (a formality that was rapidly coming obsolete). From the U.S. perspective the conflict was labeled a "police action," as if not calling it a war would serve to contain it.

Korea had all the tension of a world conflict compressed into a tiny space, threatening a massive eruption that could trigger an existential threat—the perfect symbol of the new atomic age.

The U.S. was no longer the world's sole nuclear power. The Soviet Union had successfully detonated its first plutonium bomb on August 29, 1949. And while the Soviets lacked an operational nuclear strike force at the start of the Korean conflict, it was clearly only a matter of time until they would have one.

This led to two different schools of thought. The more hawkish members of the U.S. military wanted to press the nuclear advantage while they still could and try to end communism once and for all.

More moderate U.S. leaders, including President Truman, preached restraint. ("You have got to understand that this isn't a military weapon," Truman said of the atomic bomb. "It is used to wipe out women and children and unarmed people, and not for military uses. So we have got to treat this differently from rifles and cannon.") Beyond the morally repugnant idea of using Korea as a pretext to vaporize civilians by the hundreds of thousands, if not millions, in pursuit of world domination, there was the gnawing question of what would happen if the gambit failed. The thought of a wounded Russian bear armed with nuclear weapons was terrifying.

U.S. leadership had to mull these options during a period when it was unclear who should ultimately make such decisions and what that process would be. And this played out during a rapidly escalating conflict that produced dramatic shifts in momentum. (The front basically crossed the 38th parallel four times in six months.)

In theory, America's Atomic Energy Commission (formed in 1947) maintained civilian control over the country's nuclear arsenal, which numbered somewhere between 300 and 450 atomic weapons by then. But there were clear logistical flaws in that arrangement. At some point somebody in the military chain of command had to have the authority to determine when and where to use nuclear weapons, otherwise they served no practical purpose.

Truman thought that, as commander in chief, he should have that authority. General Douglas MacArthur, who commanded not just U.S. forces in Korea but also U.N. forces, thought *he* should have that authority. And he made it clear that if he were given that authority, he would use it. At one point he requested authorization for a campaign to deploy *34* atomic weapons on targets not just in North Korea but also in China.

There's little question that MacArthur's reckless attitude toward the use of nuclear weapons played a role in Truman's decision to relieve him of his command in Korea. The irony is that, had Truman trusted his commander in Korea, he might have authorized the limited use of tactical nuclear weapons on battlefields, away from civilians. Who knows how that would have shaped U.S. policy toward nuclear weapons thereafter?

~

Both the U.S. and the Soviet Union continued to develop ever-more-powerful nuclear weapons throughout the 1950s and into the 1960s, along with more sophisticated delivery systems. In 1957 the Soviets expanded the Cold War front beyond the bounds of gravity by launching Sputnik, the first artificial satellite. Barely a decade after Hiroshima, the idea of delivering an atomic bomb via a propeller-driven airplane seemed quaint. Each side had achieved the nuclear triad, meaning they could use jet-powered bombers to deliver nuclear weapons, and also launch ballistic missiles from silos and submarines.

The individual weapons had become much more powerful, too. The arms race one-upmanship peaked with a test above the Arctic Circle on October 30, 1961, when the Soviets detonated Tsar Bomba. This hydrogen bomb was 1,570 times more powerful than the combined force of the bombs used at Hiroshima and Nagasaki. The fireball was visible for 600 miles and was capable of producing third-degree burns at a distance of 60 miles.

The scale of the arms buildup was almost impossible to process, especially when combined with the incredible pace of the space race. A question that would have seemed like pure science fiction a generation earlier was now legitimate: Could humanity manage to put a man on the moon before blowing up the Earth?

Blowing up the moon was also a possibility. The U.S. considered firing a nuke up there in the late '50s as a way of upstaging the Soviets' early progress in the space race. The mushroom cloud would have been visible from Earth, and the explosion could have permanently disfigured the man in the moon's face. (As I typed that sentence I heard the song "America [Fuck Yeah]" from *Team America: World Police* in my head.)

During the 1950s and '60s there was a pervasive sense across the U.S. that a nuclear attack was almost inevitable—but that the country could survive it by displaying that same good ol' American pluck that had pulled us through two world wars. But this time, instead of replacing butter with margarine and driving no faster than 35 mph to conserve rubber, Americans were being asked to dig backyard bunkers and stock up on canned goods and potassium iodide. Towns designated municipal fallout shelters, identified by three yellow triangles on a black circle. "In the event of an attack, the lives of those families which are not hit in a nuclear blast and fire can still be saved if they can be warned to take shelter and if that shelter is available," President Kennedy said in a televised address to the nation in 1961.

Nuclear preparedness was a constant theme during everyday discourse. The Federal Civil Defense Administration constructed "Doom Towns" at the Nevada Test Site and invited the press to document the effects of an atomic blast on a variety of houses populated with department-store mannequins. The mannequins who took shelter could survive a blast just a mile away, Americans were told, while the mannequins who blithely sat around the dinner table while the bomb dropped were blown to pieces.

See? Survival was a simple matter of choice.

Civil Defense films were also a staple in American schools. "Always remember, the flash of an atomic bomb can come at any time, no matter where you may be," one such film told the country's impressionable youngsters. "Sundays, holidays, vacation time, we must be ready every day, all the time, to do the right thing if the atomic bomb explodes."

And what was the right thing? "Duck ... and cover!"

That sing-songy advice was presented in the same tone as "Leaves of three, let it be." Just as kids could protect themselves from poison ivy by staying out of dense undergrowth, they could survive a blast of heat many times hotter than the sun's core by lying on the ground and putting their hands over their heads.

(Another way to teach your kid not to fear nukes was to buy him or her a Gilbert U-238 Atomic Energy Laboratory, which contained actual radioactive uranium ore. One of the fun activities kids were encouraged to pursue was to hide the ore somewhere in the house and see if their friends could find it with a Geiger counter.)

The 1964 movie *Dr. Strangelove or: How I learned to Stop Worrying and Love the Bomb* summed up the zeitgeist. Stanley Kubrick's film about a rogue U.S. Air Force General who orders a nuclear first strike against the Soviet Union was supposed to be a comedy. But it was hard to laugh at a doomsday scenario that seemed all too plausible. At the time, many U.S. military leaders really *were* spoiling for a nuclear exchange with the Soviets, even though it would have meant sacrificing Washington and New York, along with many other cities throughout the U.S. and Europe. But as long as we got the upper hand in the exchange, some U.S. leaders deemed that an acceptable price. "At the end of the war," said General Thomas Power, "if there are two Americans and one Russian left alive, we win."

Just to be clear: He was a real general and not a character in *Dr. Strangelove.*

~

Today there is an almost universal belief that the only value of a nuclear arsenal is deterrence. And because this has been the conventional wisdom for decades now, it's hard to comprehend just how cavalier an attitude many Americans—especially members of the military establishment—had toward nuclear weapons in the early years of the atomic age.

According to a Wikipedia page, there have been 73 nuclear accidents in the military worldwide since the days of the Manhattan Project, or roughly one every year. But only three have occurred since 1988, and two of those were in Russia. The majority of U.S. military accidents involving nuclear weapons occurred before 1970, many due to astonishing carelessness.

Consider what happened on May 21, 1946, when an "accidental criticality" occurred due to a "momentary slip of a screwdriver."

Louis Slotin, a physicist at the Los Alamos National Laboratory in New Mexico, was conducting a demonstration of how to achieve critical mass in an atom bomb by encasing a plutonium core within two halves of a beryllium shell. But Slotin, who had helped develop the Little Boy bomb used on Hiroshima, wasn't using a model. He was using the real thing. His plan was to lower the top half of the beryllium shell over the core, but to stop just short of touching the lower half. This would create a low-grade reaction that the others present could observe and measure.

The only thing that prevented the top half from touching the bottom half was a screwdriver that Slotin wedged between them. But the screwdriver slipped out. The shell closed, and the core reached criticality, emitting a burst of heat and a flash of blue light. Slotin quickly pried the top half of the shell off, but it was too late. He'd received a massive dose of radiation that killed him within days and likely led to the premature deaths of three others who were in the room.

Here are some other causes of nuclear accidents that jump out from the Wikipedia list: "A miscalculation resulted in the explosion being over twice as large as predicted ... Operator error led to a partial core meltdown ... A fire began in a theoretically fireproof area inside the plutonium processing building ... Technicians mistakenly overheated Windscale Pile No. 1 ... During chemical purification, a critical mass of a plutonium solution was accidentally assembled at Los

Alamos National Laboratory ... Accidental criticality, steam explosion, 3 fatalities, release of fission products ... Accidental venting of underground nuclear test ... Accidental loss and recovery of thermonuclear bombs ..."

About that last one: The U.S. has accidentally dropped nuclear bombs on itself with frightening regularity. It's happened in California, Florida, Georgia, Indiana, Kentucky, Louisiana, Maryland, New Mexico, North Carolina, Ohio, South Carolina, and Texas. (It's also happened in Greenland, Spain, and two Canadian provinces.) In addition, we've lost nuclear weapons in the Atlantic, the Pacific, and the Mediterranean, some of which have never been recovered.

None of those lost nuclear weapons was fully armed, of course, or the resulting detonations would have changed the course of history in ways we can't extrapolate. What if we had interpreted an accidental nuclear explosion in the U.S. as a Soviet attack and launched a full-scale retaliation?

Some of those lost nukes actually detonated with conventional explosives and scattered radioactive material over a wide area, like dirty bombs. Others came close to reaching critical mass. When a B-52 bomber suffered a structural failure over North Carolina on January 23, 1961, a pair of Mark 39 hydrogen bombs fell on the countryside near Goldsboro, a town of almost 30,000. On one bomb, all but one of the so-called failsafe devices actually failed during the plane's violent disintegration. In a 1969 memo about the incident, nuclear engineer Parker F. Jones wrote that "one simple, dynamo-technology low voltage switch stood between the United States and a major catastrophe." The switch worked—but Jones said it was well within the range of possibility that the trauma of the plane's rapid disintegration could have disabled that last link in the chain, too.

The bomb was found in a tree, dangling by a parachute. The other weapon's chute didn't deploy, and the bomb broke apart when it hit the ground at 700 miles an hour. The uranium it contained is still buried more than 50 feet below ground. (Maybe some children of the 1950s who saved their Geiger counters from the Gilbert U-238 Atomic Energy Laboratory can go look for it.)

~

And that's just a brief rundown of simple accidents. The greatest threats to humanity resulted from misunderstandings, miscommunications, and false alarms.

The most notorious incident was the Cuban missile crisis, which played out over 13 days in October 1962. The precipitating event was not an accident but an act of aggression: The Soviet Union began building nuclear missile installations in Cuba, which had joined the communist bloc under Fidel Castro in 1959. (Of course, in recounting the crisis, many Americans conveniently overlook the fact that the Soviets wanted missiles in Cuba as a check against U.S. missiles already in Turkey and Italy.) When American intelligence presented President John F. Kennedy with evidence of the missile sites, Kennedy had to quickly come up with a plan to convince the Soviets to remove the missiles while avoiding a military confrontation that might have provoked them to *use* the missiles.

That alone qualified as an unprecedented threat to U.S. security. But what made the Cuban missile crisis "the most dangerous moment in human history," in the words of historian and speechwriter Arthur M. Schlesinger Jr., was the fluidity and unpredictability of the situation, combined with a compressed timeframe. Kennedy had to do *something*, and he had to do it fast.

Although both Kennedy and his Soviet counterpart, Nikita Khrushchev, desperately wanted to avoid being responsible for sending the world back to the Stone Age, each felt at times that they might not be able to keep a nuclear conflagration from igniting spontaneously. As Michael Dobbs lays out in chilling detail in his book *One Minute to Midnight*, events came to a head on October 27, 1962, also known as Black Saturday. To begin with, both Kennedy and Khrushchev had advisers pushing them to respond to the crisis with military force. Just as Harry Truman had a bellicose World War II veteran, Douglas MacArthur, lobbying to expand the Korean conflict into China, and to use nukes if necessary, Kennedy had World War II veteran Curtis LeMay, the U.S. Air Force Chief of Staff, trying to badger him into launching a full-scale invasion of Cuba—Soviet nukes be damned. (Unknown to U.S. authorities, the Soviets had tactical nuclear missiles trained on the American base at Guantanamo Bay from just 15 miles away. Had Kennedy followed LeMay's advice and launched an invasion of Cuba, Guantanamo would almost certainly have been vaporized.)

Meanwhile, on the Soviet side, Khrushchev was trying to manage a fanatical revolutionary, Castro, who was prepared to martyr himself for the communist cause—even if it meant his entire country would be reduced to a charcoal briquette.

Kennedy opted for a naval blockade of Cuba that he hoped would buy time to negotiate a peaceful end to the crisis while also signaling America's willingness to use force if necessary. But on Black Saturday, a chain of events almost caused full-scale war to erupt without Kennedy's or Khrushchev's knowledge, much less their authorization. Here's what happened, in rapid succession:

- Soviet troops in Cuba used a surface-to-air missile to shoot down a U-2 spy plane flying a reconnaissance mission, killing pilot Rudolf Anderson. Khrushchev had authorized troops in Cuba to fire in self-defense. Taking down a plane that was shooting pictures, not missiles, was not what he had in mind. Meanwhile, Castro ordered Cuban troops to fire on any other American planes that entered Cuban airspace.

- In what seems in hindsight a remarkably poor decision, the U.S. went ahead with a planned nuclear test, code-named Calamity, by detonating an 800-kiloton bomb on an atoll in the Pacific.

- At almost the same moment as that test, a U-2 spy plane flying what was supposed to have been a routine mission near the North Pole inadvertently entered Soviet airspace. Disoriented by the strange geography at the top of the world, pilot Chuck Maultsby had wandered about 1,000 miles off course. Soviet commanders, fearful that the plane was conducting reconnaissance in advance of a nuclear attack, deployed six Soviet MiG fighters with orders to shoot the plane down. The U.S. countered by sending two F-102 fighter interceptors to fly to the U-2's defense. As part of its DEFCON-3 protocol, U.S. defenses had replaced the F-102s' conventional weapons with missiles tipped with tactical nuclear warheads. Although pilots were supposed to get authorization directly from President Kennedy to fire the missiles, there was nothing to prevent a pilot from pulling the trigger in the heat of a dogfight. Kennedy initially knew nothing about all of this. When he finally learned what was happening, his response summed up the inherent danger of having a nuclear arsenal deployed around the world, ready at a moment's notice, manned by fallible human beings: "There's always some son of a bitch who doesn't get the word."

- Maultsby, the poor son of a bitch of the moment, managed to stay just out of range of the MiGs until he cleared Soviet airspace. But then he ran out of fuel. Only the U-2's amazing capability as a glider allowed him to reach a military radar station on a desolate spit of ice. The F-102s escorted him safely back to earth.

• While all that was happening, U.S. Navy ships, enforcing the blockade around Cuba, detected the approach of Soviet subs. The U.S. had communicated to Moscow that it intended to drop practice depth charges as a signal to Soviet subs that they should surface and identify themselves or be subject to attack. But the sons of bitches commanding the Soviet subs never got the word. For all they knew, the nonlethal, state-your-intentions depth charges marked the start of World War III. The sub commanders debated whether they should respond by firing nuclear-tipped torpedoes.

Again: All of that happened in one day, within a matter of hours.

In the end, Kennedy and Khrushchev resolved the crisis by doing about the only thing they could to preserve humanity: The U.S. agreed (albeit secretly) to remove its missiles from Turkey in exchange for the Soviets removing their missiles from Cuba.

~

The Cuban missile crisis was not the only time a superpower had to quickly decide whether to respond to an imminent threat (or at least the realistic appearance of one) with nuclear hellfire. In fact, there have been a few instances when people had just minutes to decide.

In 1979, a "training scenario" was uploaded to a strategic-defense computer by mistake, triggering an alert that more than 2,000 Soviet missiles were headed for the United States. Had the threat been real, the U.S. would have had less than 10 minutes to launch a counterstrike. It took about seven minutes to trace the source of the error. We kept all our missiles snug in their silos, with three whole minutes to spare.

In September 1983, during a period of high tension between the U.S. and the Soviet Union, Soviet early-warning systems received a report of five incoming U.S. intercontinental ballistic missiles. This time it was the Soviets who had only minutes to decide whether the threat was a false alarm. Adding to the pressure, the episode occurred just weeks after Soviet fighters had shot down a South Korean passenger airliner that had wandered into Soviet airspace, sparking outrage in the U.S. and demands for a response. In the end, Soviet lieutenant colonel Stanislav Petrov concluded that the alert had to be a mistake, in part because an all-out nuclear attack would consist of more than five missiles. He was right; the detection error was due to a rare combination of atmospheric phenomena.

~

When Americans reflect on those Cold War close calls— if they reflect on them at all—they generally do so with an attitude of *Thank god those days are over.* A 2018 *Boston Globe* article about an underground bunker on Nantucket, allegedly built for President Kennedy, encapsulated this attitude. The kitschy-looking bunker, the article stated, "illustrates an interesting chapter of American history, when the threat of nuclear war was all too real, and fallout shelters were not uncommon."

Yes, the day-to-day threat of nuclear war might have been higher in the 1960s and created an atmosphere of tension and fear. But the implication that the danger has now passed is frightening on a different level.

Nuclear weapons are now largely out of sight and out of mind. But they are not out of service. So why has the fear of nuclear war receded from America's consciousness? It may be in part because there have been no superpower flashpoints as volatile as Korea and Cuba in recent decades. Now, when nuclear weapons penetrate the news cycle, it's generally because of "rogue" states like Iran and North Korea. And while it's unsettling to learn that North Korea might now be able to reach just about anywhere on the U.S. mainland with a nuclear missile, it's nowhere near as unnerving as the days when the Soviets had thousands of nuclear warheads trained on us 24/7.

But what some people seem to have forgotten is that those days have never ended. Just because the Soviet Union dissolved and "Russia" returned doesn't mean that the world has turned back the clock to the days of Nicholas and Alexandra. Russia still has about 4,000 nuclear warheads. So do we. And in early 2018, Russian president Vladimir Putin announced that his country was developing a new line of nuclear weapons that could penetrate any defense system.

That brings us to another disturbing reality. If a 21st century equivalent of the Cuban missile crisis were to erupt, there would be no JFK or Nikita Khrushchev around to defuse it. Instead, the fate of the world would lie in the hands of Vladimir Putin and Donald Trump.

That thought scares the hell out of me. And yet my sense is that few Americans have given that possibility serious consideration. (Certainly no one who voted for Donald Trump did.)

Why have we become so complacent? Is it because war itself has become an abstraction to most Americans? We haven't had a military draft in more than 45 years. Fighting in the service of the USA is now something that only a select few sign up for, and the battles take place in distant lands we know practically nothing about, for reasons we don't understand or bother to investigate.

Fewer and fewer people who lived through World War II, either as soldiers or citizens, remain each year. The collective memory of what it was like to have war envelope our entire civilization, to touch every person on a visceral level, to destroy cities and disrupt the lives of tens of millions of people throughout the First World, is fading. Too few of us have any notion of how awful the last world war was—and how much worse the next one will be.

For all the snark and condescension we direct at the "duck and cover" days, at least people back then had a realistic sense of the danger they faced. The nuclear arms buildup was never far from people's minds. There were 1,021 planned nuclear explosions at the Nevada Test Site alone. Before the Limited Test Ban Treaty ended above-ground testing in 1963, the Las Vegas chamber of commerce printed schedules advertising when tests would occur. Vegas hotels took full advantage, hosting viewing parties and "Miss Atomic" pageants. Hard to think of a more fitting response to the times. Drink and gamble all night in a burgeoning neon paradise, then have a nightcap at dawn while watching a distant mushroom cloud boil into the desert sky.

The U.S. hasn't conducted a nuclear test since 1992. But many of those ancient nukes still exist—and their age is showing. Rachel Maddow documents this in disturbing detail in *Drift: The Unmooring of American Military Power*. Her portrait of America's aging nuclear arsenal evokes Havana's preserved-in-amber car culture, in which more than 60,000 pre-1959 (and pre-embargo) American cars remain in service throughout Cuba, many kept alive with cannibalized, ill-fitting spare parts.

But in the case of our doomsday armaments, instead of trying to retrofit a Russian diesel engine into a 1957 Chevy Bel Air, we're trying to cobble together replacement Mk21 fuzes for ICBMs because our technicians have forgotten how to make them. (After reading about this in *Drift*, I came across the following online job posting from BAE Systems, a British company working with the U.S. Air Force, for its "Fuze replacement Program" (sic): "The Fuze program is replacing the Mk21 Fuze to ensure it's [sic] long term reliability for both the Minuteman III ICBM and its replacement currently known as the Ground-Based Strategic Deterrent. Mk21 Fuze is currently in the development phase of acquisition and candidates must be able to provide insight and advice to the USAF regarding contractor efforts.")

Maddow also documented the Navy's inability to reproduce "Fogbank," a secret sauce that was a critical component in the Trident submarine's W76 thermonuclear warheads, built between 1978 and 1987. Someone came up with the reasonable idea that it might be smart to whip up some fresh Fogbank after several decades, just to be sure our W76s retain their potency. The problem was, as a government report noted, that the Department of Energy no longer knew how to make Fogbank, "because it had kept few records of the material when it was made in the 1980s and almost all staff with expertise on production had retired or left the agency."

As Maddow writes, incredulously: "The experts were gone. And nobody had bothered to write anything down!"

I'm guessing that the absence of written records was a conscious choice, not an inexcusable oversight. The fewer documents you leave lying around, the less the chance of someone stealing your secrets. And while that makes sense in the moment, it's another example of how shortsighted we tend to be when it comes to technological solutions, even those with the potential for long-term consequences that we can barely imagine.

Regardless of the reason, the hemorrhaging of the brightest people from our nuclear weapons program is a major cause for concern. It's also completely understandable. If you are technically brilliant, and you are a genius when it comes to building weapons, today you would gravitate toward drones and remote-controlled tanks and other digital trickery because that's where the action is. Small, quick, surgical strikes are all the rage. The best military personnel can use a drone to take out 20 Taliban terrorists in a remote part of Afghanistan, or execute a daring nighttime raid at a compound in Abbottabad to kill Osama bin Laden.

Bright, ambitious military minds wouldn't be stimulated by the idea of serving as custodians for a bunch of oafishly large Cold War era weapons whose sheer destructive power essentially precludes their use. You can't even test the damn things anymore. And the childish truth is that engineers who work on bombs eventually lose interest if they never actually get to blow something up. Even if it's just a remote atoll in the Pacific or a house full of mannequins in the Nevada desert.

So we're left with a last line of defense that we're not sure will even work if Armageddon comes. In yet another cold-sweat anecdote, Maddow notes that it took the 2nd Bomb Wing at Louisiana's Barksdale Air Force Base 30 hours to successfully load cruise missiles onto a Stratofortress bomber and make it combat-ready. That performance would be a tad less than optimum if Vladimir Putin ever decides to unleash his new state-of-the-art nukes against us.

Even more worrisome, however, is the thought that some of our long-in-the-tooth nuclear weapons might detonate spontaneously. Just mull over this quote in *Drift* from an unnamed official, speaking during the developmental phase of America's atomic weapons program: "Nuclear weapons, even when sitting on a shelf, are chemistry experiments. They are constantly changing from chemical reactions inside of them."

The truth is, nobody really knows how stable (or unstable) our aging nuclear stockpile is. One day we might wake up to discover that North Dakota no longer exists.

And so far all I've talked about is the danger from fissile material developed for military use. I haven't even mentioned the hazards that America's 98 nuclear reactors present. Beyond the ever-present possibility of partial meltdowns like those at Three Mile Island, Chernobyl, and Fukushima, there's the urgent question of what to do with spent nuclear fuel. In yet another example of our incredible shortsightedness when it comes to any cool new technology, we failed to come up with a workable plan for disposing of radioactive waste *before* allowing all those reactors to come online.

Perspective: America is slightly more than 240 years old. That's 1% of the half-life of plutonium-239, which is used in many nuclear reactors.

The best idea anyone has come up with was to bury the stuff in the most remote, godforsaken corner of the country. The Department of Energy determined that Yucca Mountain, a desert wasteland alongside the already-contaminated Nevada Test Site, was the best candidate. The problem is that even the most remote, godforsaken wasteland sits in someone's political district. In the case of Yucca Mountain, that district belonged to Nevada Senator (and onetime Senate Majority Leader) Harry Reid. He effectively killed the Yucca Mountain proposal in 2008, with support from presidential candidate Barack Obama. As a result, most of the radioactive waste generated by those 98 nuclear reactors just stays where it is, sealed in steel-and-concrete casks, like jugs of used motor oil in somebody's garage. We'll just have to cross our fingers and hope that nothing disturbs it for the next 240 centuries.

Nuclear capability is humanity's inoperable tumor. It formed just 75 years ago—a snap of the fingers in human history—and quickly metastasized. It's the one problem we can never solve. There's no getting rid of it. There's no treating it. There's no reversing it. There's no way to unknow what we know about it. Doomsday technology exists ever after, until somebody decides to use it for its intended purpose.

CHAPTER FIVE

Not Intended for Outdoor Use

"Arrrggghhh! Nature! It's all over me! Get it off!"
—*Melman the Giraffe, in* Madagascar

The older I get, the more I feel like I'm living in a post-natural world. If I don't want to deal with the weather, I don't have to. I can just stay in my climate-controlled house. I work at home and can handle just about every routine errand remotely—banking, shopping, etc. I can import sophisticated entertainment, such as movies, music, and books. I can enjoy elaborate meals from faraway places. I can get adequate exercise using much of the same equipment I could use in a fitness club. In theory I could lead a physically and intellectually stimulating existence without ever setting foot outside again. My sense is that in just another generation or two, many people will try to live that way.

Not me. I like venturing out of the house every day to reassure myself that there is an entire vibrant planet outside my door. And after all this time I still know embarrassingly little about it.

~

Once, about 15 years ago, a tiny inhabitant of our vibrant planet buzzed my wife and me the moment we stepped out our front door. We heard a flutter and saw a flurry of feathers disappearing into the dusk. A bird had built a nest on top of the porch light, apparently not realizing the light was affixed to a much bigger nest—ours.

Over the next day or two I watched through the living room window as the bird came and went. I couldn't tell what it was. It was smallish like a sparrow or chickadee, but it didn't look quite like either.

I noticed that the bird often pumped its tail while perched on the porch railing. I didn't think that was unusual. Don't all little birds do that?

No, they don't. Some rudimentary internet research revealed that that's a distinctive feature of the eastern phoebe, a common flycatcher. An accompanying photo confirmed that we had positively identified the intruder. Once I learned to recognize this new (to me) species, I discovered just how common it is in New Hampshire. I started seeing phoebes everywhere. And hearing them. Supposedly they got the name "phoebe" because that's what their call sounds like—although to my ears the chickadee's call actually sounds more like *fee-bee* than the phoebe's. I think a phoebe's call sounds closer to a whisper, or a kid trying to learn how to whistle.

My wife and I were happy when we noticed a pair of phoebes rebuilding the nest on our porch light the next spring. (Phoebes often return to the same site each year.) Still, it was disruptive, for the birds and for us, to have the nest so close to the front door. We couldn't avoid disturbing them whenever we came or went. So before the next breeding season I built a small wooden platform in the far corner of the porch, hoping the phoebes would take the hint and build there instead.

They did. I'm not sure which generation we're on now, but we look forward to the phoebes returning to our front porch every spring.

~

It took that first little feathered trespasser to open my eyes to the wide variety of birds in our neighborhood. Since then I've made an effort to tell them apart. Some, like the scarlet tanager and indigo bunting, are so exotic-looking that I'm amazed I never noticed them before. I had recognized robins and blue jays and not much else. I know my birds a lot better now, but I'm still appallingly weak at identifying many other elements of the natural world, including most plants.

It isn't that I'm not observant or curious. It's just that for most of my life I've paid far more attention to the man-made features in my environment than the natural ones. I grew up on a busy state highway, and the kids in our neighborhood would sometimes gather on our front porch to see who could be the first to identify each passing car. We could distinguish models that looked almost identical. (The 1968 Camaro had side marker lights, the 1967 Camaro didn't.) But the difference between a purple martin and a grackle? Forget it.

And it's not as if we weren't exposed to nature. We played in the woods all the time. We just had no information about what we were seeing, or any incentive to look for it. Because, unlike the vast majority of human beings who preceded us on this planet, we didn't have to learn how to interact with nature to survive.

Meanwhile we were inundated with information about cars and gadgets. ("Make hundreds of julienne fries in seconds! Chop onions without shedding a single tear! Grate cheese or mounds of carrot salad!") Technology was becoming more integral to popular entertainment—starting with the locus of every living room, the television set. Drive-in restaurants and movies proliferated. More and more toys ran on batteries or had to be plugged in. Electric guitars transformed music. (I remember strumming a badminton racket as I pretended to play along with the Beatles' "Can't Buy Me Love," the first rock-and-roll 45 in our house.) As children of the 20th century, our interests inevitably gravitated toward interactions with technology rather than with nature.

Even in situations where technology and nature intersected, technology got the upper hand. Fishing, for instance. My friends and I all learned to identify the handful of freshwater species hardy enough to survive in the rivers and streams where we lived. (*Freshwater*, in this case, was a misnomer if ever there was one.) We were more like refugees of an environmental apocalypse than avid outdoorsmen. One local creek was known as Stink Run because of the warm industrial discharge it contained. It never froze, even on the coldest winter days. If you got antsy waiting for the official start of fishing season in April, you could go to Stink Run in February or March and get the kinks out. Using kernels of frozen corn for bait, you could pluck a few pathetic little bullheads or yellow perch from the steaming water.

Our favorite fishing spot was where a natural gas pipeline crossed the Allegheny River just downstream from a sewage treatment plant. The first fish I ever caught there was a sucker, which is as unappealing as it sounds. Using its downward-facing mouth (which is encircled by a ring of gristle), it drifts along the bottom Hoovering up the muck. We also pulled in a lot of carp, which are basically overgrown goldfish. Catching them was easy. You just put a night crawler—or even a portion of one—on your hook, cast slightly upstream, and let your sinker serve as an anchor. Then you propped your pole up against the pipeline and watched for the telltale bend at the tip.

As we got a little older, we started going after sport fish, like largemouth bass, walleye, and muskellunge. Catching those required a more sophisticated approach—which, of course, involved technology. We graduated from the clumsy Zebco 202 to sleek, open-face spinning reels. And we experimented with different kinds of seductively named artificial lures. My favorite was the Mepps Black Fury, a spinning lure. One of the first times I used one, I hooked a small muskie on a rainy evening. From then on I was hooked on the Black Fury, too.

A neighbor had uncanny success catching walleyes using Red Devil spoons. I could never duplicate it. I tried, but my efforts were halfhearted. Part of my problem was that I couldn't comprehend how a Red Devil spoon worked. With a Black Fury spinner, I could understand how a dark, rotating blade could trick a dim-witted bass or muskie into attacking it. In the murky river it probably looked sort of like a baitfish swimming past.

But the lure of the Red Devil lure eluded me. It didn't look like it was swimming at all. It fluttered in the water like tickertape. And it didn't remotely resemble any living creature I had ever seen. It looked like a large, legless, headless, candy-striped beetle with a grappling hook hanging out its ass. I remember holding one up and wondering aloud, "If you were a fish, why would you want to eat *this*?"

"OK, smart guy," a fish might well have responded, holding up a frosted cherry Pop-Tart, "why would any human want to eat *this*?" (Of course, to hold up a Pop-Tart, a fish would have to have hands. But it wouldn't surprise me if some of the fish in Stink Run did.)

Touché, mon poisson à deux mains.

~

I wish I could travel back in time and spend a day in the 1800s. My goal wouldn't be to witness something historic, like the beginning of the Lewis and Clark Expedition or Lee's surrender at Appomattox. I would just like to see what an ordinary day was like back then. I would like to breathe nineteenth century air and study the quality of light.

What I would most like to do, though, would be to sample the food and drink to see how they compared to today's.

I'm guessing a meal from the 1800s would make me violently ill. Not necessarily because the food would be spoiled (although that's a distinct possibility), but because it would be so different from 21st century processed food that it would shock my system. (In a Nautilus article subtitled "Burgers and fries have nearly killed our ancestral microbiome," science journalist Moises Velasquez–Manoff reports on recent studies suggesting that "by failing to adequately nourish key microbes, the Western diet may also be starving them out of existence.")

Just think about the changes since the 19th century in something as basic as drinking water. Civilizations have experimented with various approaches to water treatment since before the days of the Roman aqueducts. Boiling, sand filtration, even charcoal—people tried 'em all in the long-ago, mostly in an effort to improve the taste. (Hard to believe, but a consensus on the dangers of waterborne contaminants didn't emerge until fairly recently.) Even so, people drank water straight out of the ground during America's westward expansion in the 19th century.

During my notional day trip to the 1800s, I suppose I would do the same. I'm curious to know what water tasted like it in its natural, pre-industrial state. Still, drinking it would feel like a suicidally *un*natural thing to do. I think I would have preferred well water to dipping from a spring or stream. At least I wouldn't have seen the source. Consuming water that had just splashed over the slimy rocks right in front of me? That would require overriding an alarm system that had been activated in childhood. Everyone knows the only safe drinking water is the stuff that comes from a faucet (and even then you should put it through a home filtration system). Either that or from a plastic bottle.

We can trace this modern conception of safe drinking water to a man named John L. Leal. In 1908, at Leal's urging, Jersey City, New Jersey became the first municipality to treat its water supply with chlorine. (Of *course* that practice started in New Jersey.) At the time, there were no laws regarding water quality; the first federal regulations didn't go into effect until 1914. Since then we've made up for lost time. The most recent iteration of the Safe Drinking Water Act (originally passed in 1974) runs more than 1,000 pages. It's contaminated with sentences like this:

(B) ESTABLISHMENT OF LEVEL—If the Administrator establishes a maximum contaminant level or levels or requires the use of treatment techniques for any contaminant or contaminants pursuant to the authority of this paragraph—(i) the level or levels or treatment techniques shall minimize the overall risk of adverse health effects by balancing the risk from the contaminant and the risk from other contaminants the concentrations of which may be affected by the use of a treatment technique or process that would be employed to attain the maximum contaminant level or levels; and (ii) the combination of technology, treatment techniques, or other means required to meet the level or levels shall not be more stringent than is feasible (as defined in paragraph (4)(D)).

Read enough of the Safe Drinking Water Act, and you could reasonably conclude that water is a type of poison.

~

Food has changed even more dramatically than water since the 1800s. Almost everything about it—the way it's grown, the way it's prepared, the way it's stored, the way it's cooked—is different.

But at least food in the 1800s was *food*. It wasn't a chemistry experiment, like so much of what's on supermarket shelves today. Not that all 19th century food was good—particularly not among the pioneers. Their staples along the trail included hardtack and corn mush.

Hardtack was a cross between a cracker and a biscuit. It was made with flour, water, and sometimes a bit of salt. It kept a long time—though not always in perfect freshness. (There's a Civil War memoir called *Hardtack and Coffee*, a title that has a quaint appeal—until you realize that it's based on the practice of soldiers breaking their hardtack in coffee in order to flush out the weevils, which were "easily skimmed off, and left no distinctive flavor behind.")

Once the homesteaders staked their claims (as close to a source of drinking water as they could get), they settled in and became what we would call locavores today. They grew their own vegetables—corn, mostly, because they could grind it into meal to make cornbread and johnnycakes. Beans were another versatile plains favorite.

Homesteaders also hunted game ranging from bison and elk to rabbits and squirrels, depending on where they were. They might round out their diet with turtle, muskrat, raccoon, and possum. (In general, cows were a source of milk, not beef.) "It took persistent, resourceful people who ate whatever was available and weren't overly particular about how things tasted," Joe Johnston of *Wild West* magazine wrote on HistoryNet. "Spices were sparse, and cooking was always an adventure in ingredient substitution."

Cities back east had restaurants where you could order meals similar to what you could get today. During the 1850s, the breakfast offerings at Cincinnati's Burnet House included ham and eggs, waffles, scrambled eggs, bacon—even "an omelet with fine herbs." For dinner (served between 1:00 p.m. and 3:30 p.m., which was typical of the time) you could have had various cuts of steak, fresh fish, or pork chops, along with potatoes or rice. (On the other hand, if you wanted the full 19th century immersion, you could have tried the fricasseed tripe.)

Not a word on the menu about any salads, fruits, or vegetables.

The biggest challenge to providing fresh food at the time was the lack of refrigeration. People resorted to salting, smoking, and pickling meats; vegetables could be dried or stored in root cellars, and some fruits could be dried or canned. The seasons dictated what people ate to a much greater degree than they do today.

Advances in refrigeration—first in railroad cars in the 19th century, and by the 1920s in households—made it possible for people to eat fresh fruits and vegetables from all over, almost all year round. That should have put America on the fast track to better health. We should all look like Tom Brady or Gwyneth Paltrow by now.

Fat chance.

~

The advent of refrigeration coincided with another development in American nutrition: an explosion of packaged, processed convenience foods.

This has had an outsize impact on American culture. And it has resulted in a lot of outsize Americans. Here are eight milestones along the wrapper-strewn path to poor nutrition, viewed through the prism of my own dubious dietary history:

1896: Colonel John Pemberton invents Coca-Cola.

1897: A Campbell Soup Company chemist named John T.

Dorrance devises a formula for five varieties of canned, condensed soup, including tomato.

1905: Lombardi's, America's first pizzeria, opens in Manhattan.

1906: The Battle Creek Toasted Corn Flake Company introduces a new variety of cornflakes that contain sugar.

1916: Piggly Wiggly opens the first self-service grocery store.

1921: The Taggart Baking Company of Indianapolis unveils Wonder Bread.

1924: Clarence Birdseye develops the quick-freezing method that spawns the frozen food industry.

1955: Ray Croc launches McDonald's System, Inc.

It's uncanny the degree to which those eight developments shaped my diet when I was a kid. For breakfast, I almost always had processed cereal dosed with tooth-decaying quantities of sugar. By the 1960s and '70s food manufacturers were peddling candy for breakfast, and they didn't pretend otherwise. The names said it all: Sugar Smacks, Cocoa Puffs, Sugar Pops, Cocoa Krispies, Sugar Frosted Flakes, Count Chocula, Sugar Coated Rice Krinkles. One cynical variation was Froot Loops, a concoction of flour, sugar, and garish dyes—and not a speck of actual fruit. But it *sounded* healthy.

Those cereals were as meticulously engineered as fishing lures, packaged into bright, loud TV commercials, and cast into the polluted stream of morning cartoon shows. The target audience was obvious. "Trix are for kids!" No shit.

Occasionally I would take a break from cereal and switch

to frozen waffles topped with margarine and "table syrup," which is a euphemism for corn syrup artificially colored to look like maple syrup. Or I would have Pop-Tarts. (Unless my mom opted for the store-brand "toaster pastries." Those tasted like Pop-Tarts made with a cheaper grade of cardboard.)

When the sugar rush faded after a few hours, I would be ready for lunch. That usually meant sandwiches made with enriched white bread. What did "enriched" mean, exactly? Let's consult the World Health Organization's glossary of terms. Turns out that before we can grasp "enriched," we have to know the definition of a similar term, "fortified":

"Fortification is the practice of deliberately increasing the content of an essential micronutrient, i.e. vitamins and minerals (including trace elements) in a food, so as to improve the nutritional quality of the food supply and provide a public health benefit with minimal risk to health." This is distinct from food "enrichment," which the WHO says "is synonymous with fortification and refers to the addition of micronutrients to a food irrespective of whether the nutrients were originally in the food before processing or not."

Ah, I get it. So basically, once you've squeezed most of the nutritional value out of white bread dough by producing it in industrial quantities, you try to artificially stick some of those nutrients back in—and while you're at it, you toss in a few vitamins or minerals that were never in the basic ingredients to begin with. (I picture a technician crushing One A Day vitamins with a mortar and pestle and sprinkling the powder into the raw dough just before baking.)

What did I put on my enriched white bread sandwiches? Processed peanut butter, usually combined with jelly, jam, honey, or Marshmallow Fluff. For variety, I would sometimes switch from peanut butter to processed American cheese or thin-sliced packaged cold cuts loaded with enough salt and preservatives to have done a 19th century homesteader proud. I also vaguely recall something called Swanson Chicken

Spread. I'm tempted to Google it, but it's probably best left in the dim recesses of my memory.

I usually finished lunch with a cookie or packaged cake.

We often ate something approaching real food for dinner, like roast beef or pork chops. But it was usually supplemented with packaged foods like frozen peas, frozen corn, or a frozen "medley" (the lima beans ruined it for me), along with instant mashed potatoes, dinner rolls that came in tubes, and salad dressing in a bottle (for our iceberg lettuce). Dessert was more packaged cookies or cakes.

Other times dinner came straight from the supermarket shelf or freezer: Dinty Moore beef stew; Campbell's Chunky Sirloin Burger soup; TV dinners; individual pot pies; boil-in-bag chicken, turkey, or beef; frozen pizza. On the rare occasions when we ate out, the choice was almost always fast-food burgers or pizza.

My go-to drink was whole milk, often flavored with chocolate or strawberry Quick. I also drank a lot of soda and packaged drinks like Kool-Aid and Wyler's, and occasionally lemonade made from frozen concentrate (it was kind of a pain to make). Plain water? The only time I ever drank the stuff was when it came from the fountain at school.

I don't think I ever gave the quality of my diet any thought back then, even as I got a little older. Convenience was usually the most important consideration at meal times. By the time I was in junior high, my mom was working the night shift as a nurse and everyone in the house was on a different schedule. If I had a baseball game to get to, bagged meat product was fine. Maybe I was deluded by the Froot Loops fallacy—a "hot roast beef sandwich" sounded like a decent meal, even if it was really just a frozen pouch of compressed, salted butcher's offal that I tossed in a pan of boiling water for 10 minutes and then dumped over two slices of bright white bread that had been given a B-12 shot (another image that springs to mind when I try to picture the "enrichment" process).

~

The American diet is another example of the hubris that technology breeds. Growing healthy food and turning it into decent meals takes a lot of effort. Americans decided to outsource that process because we had better things to do with our time. Now we're paying the price. According to the World Health Organization, the U.S. ranks 31st in the world in life expectancy, just behind Costa Rica and just ahead of Cuba. Roughly seven in 10 Americans over the age of 20 are overweight, and four in 10 are obese. As of 2015, 23.4 million Americans had been diagnosed with diabetes, compared to just 1.6 million in 1958.

This is one of the more glaring ways that technological evolution and natural evolution have ended up at cross-purposes. The cravings that people feel for foods high in fat and calories are rooted in an ancient survival instinct. "Adaptive mechanisms that were meant to protect us from starvation have now, in fact, led to the dual epidemics of obesity and diabetes," Dr. Haider Warraich, who is researching a book on the history of heart disease, told *The Atlantic*.

And semi-reformed junk-foodies like me are finding that the solution isn't as simple as trying to eliminate processed foods (75% of which contain added sugar). For one thing, cooking from scratch is labor intensive. And you need to keep a lot of ingredients on hand. (Usually I cop out and use a couple of premade components.) Also, even if you think you're buying fresh ingredients at the supermarket, you often end up with just another form of processed food. How many times have you picked up a ripe red strawberry, taken a bite, and tasted ... not much of anything?

Although less noticeable to the average person (certainly the average coastal dweller), the changes on the American farm over the last century have been at least as dramatic as changes in the American kitchen, if not more so. Between 1920 and

2012 the number of farmers in the U.S. dropped from 32 million to 3.2 million. But the average farm got much bigger and more specialized, thanks largely to advances in equipment, fertilizer, and irrigation. Now, instead of serving a local market by devoting a few acres to corn, a few acres to peas, a few acres to pumpkins, and a few acres to chickens, farms devote mega-acreage to single crops for a global market. (According to the American Farm Bureau Federation, only about 8% of U.S. farms market foods locally.)

"Economies of scale" is the go-to phrase for describing this approach. The phrase you don't hear (but should) is "inferiorities of scale." Because the former often results in the latter. It's unnatural selection—when your objective is to create maximum yield as efficiently as possible, you engineer produce that sacrifices nutrients and flavor for consistency and a longer shelf life. (Or you manipulate the shelf life by shipping unripe fruit and then artificially ripening it with things like ethylene gas.)

And when your ultimate goal is profit, you choose the crops that will make the most money, not the ones that are the most healthful or environmentally sound. That explains why corn is America's top crop. Since 1983, U.S. corn production has increased by 50%, thanks in large part to the Federal Agriculture Improvement and Reform Act of 1996. According to the USDA, "The Act permitted farmers to make their own crop planting decisions based on the most profitable crop for a given year."

That would be fine if we had a truly free market. But we don't. The deck is stacked in favor of corn, which is why it's such a versatile moneymaker. Of the more than 90 million acres of corn harvested each year, less than 10% is for human consumption (direct human consumption, anyway). About 40% of U.S. corn ends up in ethanol, thanks to a taxpayer-subsidized scam of epic proportions called the Renewable Fuels Standard. (It's why 10% of the fuel you pump now contains ethanol—even though ethanol is arguably less energy efficient than the gasoline it's added to.)

Another 30–40% of U.S. corn ends up as animal feed—some in the form of something called "distillers dried grains with solubles," a nutrition-depleted byproduct left over from ethanol production. A minute portion of U.S. corn also finds its way into animal feed via the scenic route. Corn syrup and modified cornstarch are among the ingredients in Skittles, which sometimes supplements the "distillers dried grains with solubles" fed to cows. (This practice came to light when a sheriff investigating a large spill of mysterious red pellets on a Wisconsin highway detected "a distinct Skittles smell.")

I suppose it shouldn't come as a surprise that a country that hooks its kids on sugar also feeds candy to its cows. A *Fortune* article about this episode included an *Onion*-like postscript: "This story has been updated to show comment from Mars."

Of the small fraction of U.S. corn that is actually used for human consumption, much of that is in the form of high-fructose corn syrup. Each year Americans consume more than 40 pounds per capita of the stuff. And, like ethanol, high-fructose corn syrup is produced with the help of government subsidies.

Yes: At the same time many public health departments are warning you not to eat high-fructose corn syrup because it's bad for you, the government is also taking a portion of your paycheck and giving it to farmers so they can produce more high-fructose corn syrup.

Besides being harmful to public health, a corn-centric economy is bad for the environment. More and more of America's heartland is given over to corn production at the expense of healthier foods like soybeans. (America now imports 31.1% of the vegetables it eats, compared to just 5.8% in 1975, and more than half of our fruit.) Corn requires more water than the annual precipitation over much of the Great Plains is able to provide. Irrigation is leading to a gradual depletion of the massive Ogallala Aquifer. Corn has also supplanted much hardier grasslands and portions of The Great Plains Shelterbelt, the 220 million trees that were planted as a hedge against erosion in the wake of the 1930s Dust Bowl catastrophe.

It's a mistake to assume that because fugitive dust no longer boils over the plains in ugly black clouds that erosion is under control. According to Matt Hongoltz-Hetling's article "The Land of Iowa" (part of the Weather Channel's United States of Climate Change series), Iowa's layer of rich dark topsoil has diminished from a depth of 15 inches in 1900 to just seven inches today. And millions of additional tons wash away each year, carrying nitrogen-laden fertilizer into the Mississippi River watershed. All that nitrogen is expanding the dead zone in the Gulf of Mexico, which is now roughly the size of New Jersey.

As Hongoltz-Hetling writes, "Historically, the industrial Iowan approach to growing corn has been to use plows and chemicals to reduce the soil to a lifeless powder, and then to truck in the fertilizer needed to support plantings, until the leaves of corn stalks are waving feebly in the wind, like hollow victory flags."

What happens when a society places a premium on maximum technological efficiency in pursuit of the lowest common denominator? It ends up squandering its most valuable farmland on a nutrient-depleted cornfield the size of Montana.

CHAPTER SIX

Ask Your Doctor if Immortality Is Right for You

"The only thing worse than a cell that forgets how to live is one that refuses to die."

—Haider Warraich, M.D.

Sometimes I feel old and crotchety complaining about life in this technological magic show. Not to mention ungrateful. Technology's wizardry has benefited me in more ways than I could ever count. (Modern dentistry, for starters. Imagine not being able to fix a simple cavity.)

Here's the most obvious debt I owe to technology: I would be well on my way to a miserable, premature death if a routine colonoscopy screening hadn't detected early-stage cancer. More technological sleight of hand enabled a surgeon to cut out the traitorous cells robotically.

Start to finish, the process involved a series of small miracles. The tiny camera that a gastroenterologist shoved up my ass, which discovered the incipient cancer. The tattoo the gastroenterologist later applied to the affected area with India ink. (In my case, the joke about not being able to show anyone my tat was literally true.) The anesthetic that knocked me out during the seven-hour operation so I didn't feel a thing. The IV drip and the catheter that allowed my body to function normally throughout. The pain blocker that enabled me to get a night's sleep after the operation. From the next morning on I didn't even need Tylenol, let alone anything stronger.

And my experience was hardly unique.

Medical magic also gave me many extra years with my parents. My dad lived to 75, my mom to 80—almost their precise actuarial projections. (Knowing my mom, though, I'm sure she raised a stink at the Afterlife Service Desk about being shortchanged a year on her 81-year life expectancy.)

My parents lived long lives even though they both smoked for more than 50 years. They also had diets high in everything terrible and low in anything beneficial. They regarded exercise as self-inflicted torture. They stressed out over subatomic problems. (I felt like saying, "You survived Nazi bombing raids. I'm sure you'll survive a restaurant serving you a cup of tea that's five degrees cooler than the optimum temperature.")

My parents needed lots of help to make it to the finish line. My dad's medical history included ulcers, transient ischemic attacks, mitral valve surgery, bladder cancer, congestive heart failure, and kidney disease. My mom needed stents in her coronary arteries. She also had acid reflux and chronic emphysema. She had her gall bladder removed when she was 79.

Medical science's ability to prolong life is the plainest manifestation of artificial evolution. As Dr. Haider Warraich writes in *Modern Death*, "The doubling of human life expectancy from 40 to about 80 over the course of about 150 years means that a change in genetics has little role to play in the plasticity that we have demonstrated in our lifetimes."

Warraich also writes that such a spike in longevity "has been noted for no other organism in history. Not only has change of this extent, which mostly started after 1900, never been observed for any living thing outside a laboratory, it has never been achieved even with cells or organisms in an experimental setting."

Modern America *is* an experimental setting. It's as if we're experiencing a mass case of Munchausen's syndrome by proxy. We've created a society that is almost guaranteed to make people sick. (The list of things that might have caused my cancer is virtually endless. Maybe the seed was planted years ago when I was fishing in Stink Run and eating Pop-Tarts for breakfast, Fluffernutters on enriched white bread for lunch, and corn-fed fast-food burgers for dinner, washed down with sugar-saturated soda.) But then—*Voila!*—the massive medical-industrial complex can perform miraculous feats to enable all these debilitated people to stick around long enough to draw every last breath (and pay every last dollar) and drive everyone else's health insurance premiums into the exosphere.

As Warraich notes, "Death, in most cases, is no longer a sudden conflagration, but a long, drawn-out slow burn."

It certainly worked out that way for my parents. Toward the end they lived in a shadow world. Their days were physically unpleasant and spiritually unfulfilling—yet still offered enough flashes of life to string them along. Call it existential FOMO. It's another side effect of living in a society enslaved by the idea of ceaseless forward motion. There's always one more thing coming over the horizon that we want to be a part of. A birthday. A graduation. A wedding. The Super Bowl. A new novel by what's-her-name. A new movie by what's-his-face. The season finale of *Dancing with the Stars*. A trip home. A tractor pull. It could be almost anything. We just know that we don't want to miss it. We've been conditioned by a lifetime of previews and promos and coming attractions, by a media that projects potential candidates for the next presidential election as soon as the last one is over.

And the medical profession is geared toward prodding people along as long as they have a leg to stand on—and sometimes even when they don't.

~

The beginning of the end for my dad was when he had his left leg amputated. He'd been having circulation problems that had exacerbated a dangerous infection. A hotshot young surgeon thought he could fix my dad's leg with a cutting-edge vascular reconstruction technique that he was excited to try.

The surgery didn't work. My dad's leg turned gangrenous and they had to cut it off above the knee.

Afterward the surgeon insisted that the procedure actually *did* work. My dad's leg was just too far gone, he explained, for his wondrous procedure to reverse the damage. My mom dismissed Dr. Wonderful with a wave and walked away. "He didn't give a damn about Bill," she told me later. "All he cared about was the damn operation. He was like a little kid, all excited to try it so he could show everybody what he could do, and when it didn't work we were supposed to feel bad for *him*."

The spirals of my dad's decline became tighter and tighter. He had to go on dialysis because of his failing kidneys. A couple of times when I picked him up from the dialysis center he was gray and shivering and looked like he'd been plucked from a casket.

He talked about giving up dialysis. When people pointed out that he would soon die without it, his reaction was, *Yeah, so?*

By then he was also taking a cocktail of medications to control things like high blood pressure. He hated the side effects. A doctor offered to perform a procedure that might bring some relief but also carried considerable risk of stroke. My dad told him to go for it. He stroked out and died the night of the procedure. I'm sure he was OK with the outcome. It might even have been the one he was hoping for.

My mom spent her last 5½ years as a widow. Her situation was so depressing (and depressingly common) that it was easy to overlook how extraordinary her life had been. She'd spent her first years in a crowded, run-down part of Glasgow, crammed into a two-room tenement flat with five other people. Her family shared a common bathroom with several equally deprived families on the same floor. She spent her last years alone in a three-bedroom house on Cape Cod, a setting that I'm sure was way beyond her wildest childhood imaginings. She had her own car and an internet connection that allowed her to communicate with her kids scattered around the U.S. as well as her relatives back home.

I'm positive she was happier in the tenement, even when Adolf Hitler was lobbing bombs in her direction.

Loneliness was the most obvious reason my mom was so unhappy toward the end. But poor health exacerbated her misery. Emphysema was inexorably suffocating her. And yet she kept smoking. "I'm going to cut down," she'd say. Not quit—*Let's not get carried away here!*—just cut down.

Another problem was that my mom had outlived her sense of purpose. She spent the last years of her life literally not knowing where to put herself. She didn't want to stay alone in her house, but she lacked the physical and emotional wherewithal to seriously investigate any other option. My sister and I each lived within a couple of hours' drive and each of us did our best to help her make up her mind, but nothing panned out.

At one point my mom talked about moving back to Scotland. A two-month visit to Glasgow sufficiently scratched that itch. She told me she didn't want to live at my house because New Hampshire was too cold and snowy. My sister offered to add an apartment onto her home in suburban Boston, but my mom nixed that idea because suburban Boston is in suburban Boston. Or something.

At the onset of what turned out to be her final winter my mom said she couldn't face another season of silence and 4:30 sunsets. She told us she was thinking of closing the house and moving to an assisted living facility until spring. I'm sure that, left to her own devices, she wouldn't actually have done it.

So my wife and I essentially did it for her. Rather than move her to an assisted living facility on Cape Cod, we found one near our home in New Hampshire that would give her a one-month trial, at far less than it would have cost in Massachusetts. We pointed out that she would have onsite medical care (she lived in fear that she would collapse in her house due to shortness of breath and not be able to get help), she would have plenty of people her own age to socialize with, and she would have daily visits with the grandson that she so obviously adored. And she wouldn't have to worry about getting snowed in. What's not to like?

Everything, apparently. My mom approached this experiment with an attitude that could best be described as "defiant hostage." First, she did her best to sabotage it before it even started. She was supposed to make the move in January, but she postponed it until February so she could have her gall bladder removed.

There's a backstory there. For years my mom had suffered her gall bladder attacks in silence. Well, no, not in silence. Actually this 4' 11" 98-pound woman with a Glasgow accent would pace around the house after every meal belching like Barney from *The Simpsons*. But she resisted all medical advice about how to fix the problem.

Until that January, when she was supposed to make her temporary move to New Hampshire. Then suddenly she was all-in on having her gall bladder removed. And just in case anybody missed the not-so-subtle message (I'D RATHER DIE THAN GO UP THERE), she insisted on a DNR order before surgery. That was a first for the surgeon before a simple gall bladder procedure, but he humored her and went along with it.

So in February she came up to New Hampshire without her gall bladder but with plenty of attitude. I think the main source of her surliness was that the assisted living facility didn't allow smoking anywhere on the grounds. She was reduced to sneaking cigarettes on the balcony outside her room in the February cold (she didn't think I knew). She hated herself for it—and some of that hate spilled in my direction.

And yet, for all her theatrical misery with the surgery and the DNR and the complaining, my mom was no more prepared to die when the time came that spring than she'd been when it was time to move to New Hampshire that winter.

It was Memorial Day weekend. My mom had been back in her house, alone, since the assisted-living experiment had mercifully ended. I got a call from one of her neighbors. My mom hadn't answered a knock on the door. But her car was in the driveway, so the neighbor let herself in (she had a key). The neighbor found my mom slumped on the bed. She was breathing but unresponsive. The neighbor called 911.

I drove to Cape Cod not expecting to ever talk to my mom again. I spoke to the ER doctor on the phone, and she suspected that my mom had had a stroke. But she hadn't. And the next day she rallied. As she regained consciousness, she tried to pull out her intubation tube.

What did that mean? Was it just a panicked response to a gagging sensation? Or was it another indication that she didn't want to be resuscitated and kept alive by artificial means?

Over the next two days my mom recovered to the point where she could breathe without the intubation tube. That allowed her to talk with my sister and me. At first she seemed to accept that this was the end. She was stoic. She told us how angry she was at herself for all those years of smoking—something she had never admitted before. ("You'll always pay the price.") She said a few other dying-declaration-type things, too, like "You're all legitimate." (That one made us laugh.)

Before long I noticed that her breaths were getting shorter. The calm, stoic demeanor started to fade. A nurse asked if she would like a sedative. She said yes.

A doctor asked to speak privately with my sister and me. He confirmed what I suspected: My mom would never be well enough to go home. From now on she would need assistance to breathe, beyond her oxygen bottle, because her lungs couldn't adequately expel carbon dioxide. Already, in the short time that she had been without the intubation tube, her CO_2 level had begun to climb again.

The doctor asked us if my mom had ever expressed a preference about whether she would want to be kept alive with artificial means. I said she had, and told him about the DNR request before gall bladder surgery. And that was just one of many instances when my mom had said she wouldn't want to be tethered to a machine.

The doctor went to talk to my mom. He knew he needed to hear from her while she could still communicate, and before the sedative kicked in. He explained that the only way to stabilize her breathing was to put her on a ventilator permanently. His understanding was that she didn't want that. Was that correct?

Her initial answer surprised me: "Maybe you could just do it for a little while at a time."

The doctor answered in a gentle voice: "I'm afraid it doesn't work that way."

Confronted with a final, binary choice, my mom opted not to go on the ventilator. But the decision wasn't as clear-cut as she'd always assumed it would be.

She soon lapsed back into a coma. She lingered a couple of days before she died—just long enough to make it to my 52nd birthday. I'm convinced she did that on purpose, to be sure I would always remember to mark the anniversary.

~

I took that extended detour through my parents' medical histories to illustrate the moral morass we've created with all these sophisticated life-extending technologies. No two diseases, no two doctors, and no two patients are exactly the same. And yet we try to throw a blanket ideology over all of medical science. It can't possibly work that way. You have to consider each life, and each death, on a case-by-case basis.

Just look at my parents' circumstances. In all my conversations with them, they expressed the same opinion about being kept alive by artificial means. Neither of them wanted that. They were adamant about it.

At the end my mom seemed to waver. But I think she was ambivalent about going back on the ventilator not because she feared dying per se but because she feared dying of suffocation. It's a nasty way to go. (In the end I think she was sufficiently sedated to be unaware of what was happening. I certainly hope that was the case.) Contrast her situation with my dad's. He showed no ambivalence at all. He opted for a risky procedure knowing he would either wake up feeling better or not wake up at all.

So while neither of my parents wanted to keep living in misery, it just so happened that my dad was presented with an escape route free of discomfort and my mom wasn't.

For another contrast, look at the bedside manner of two of the doctors involved in the late-stage decision-making. My mom's ICU attending struck just the right tone for the circumstances. As for Dr. Wonderful—I doubt he saw any difference between a living patient and a cadaver. The human body was just a medium that he could he practice his wonderfulness on.

My parents' circumstances weren't extreme, either. As end-of-life scenarios go, theirs were actually pretty benign. They didn't linger for years in nursing homes. We didn't have to watch them endure metastatic cancer, with radiation or chemo treatments that can be worse than the disease. Neither suffered from dementia or Alzheimer's. I recognized my mom and dad as my mom and dad right through my last moments with them. Millions of others aren't so fortunate.

Nor were my parents denied access to potential lifesaving treatments because of where they lived, because of limited insurance coverage, or because their names were too far down on a transplant list.

In the context of 21st century medicine, my parents' deaths were perfectly normal. But in the context of history, 21st century deaths—like just about every other aspect of life in this century, as I noted in this book's introduction—are *not* normal. Dr. Warraich wrote a book to try to drive that point home. "After remaining more or less static for many millennia," he notes, "death changed on a fundamental level over the course of a century. Modern death is nothing like what death was even a few decades ago."

Our current quandary with medical ethics is yet another consequence of living in a technology-centric society that is obsessed with continual improvement. We start to assume that technology can fix *anything*. (COPD? There's an app for that.) TV reinforces this attitude with all those commercials pushing prescription drugs (the U.S. is one of the few countries that allow them). In a single evening you might hear about drugs that can cure everything from diabetes to depression to fibromyalgia to ulcerative colitis to erectile dysfunction. Those ubiquitous commercials also condition us to dismiss potential side effects as just so much lawyerese that we shouldn't concern ourselves with. (*While using prescription Runphromdeth, consult your doctor if you experience ingrown elbows, furry stools, a narrowing of the ears, farts that smell like North Dakota, or the constant feeling that you're about to sneeze even though you never do ...*)

The main problem with our culture of limitless cures—beyond the incredible expense, which threatens to capsize the economy—is that it produces diminishing returns with no tidy endpoint. Think about that doubling of life expectancy, from 40 years to 80. That would be an unequivocal improvement in the human condition if the second 40 years were as good as the first. But they aren't. People decline—particularly in that last decade en route to 80.

Early on in those extra 40 years, most decisions about whether to treat health problems are pretty straightforward. I was 58 when I got my cancer diagnosis—almost halfway through my 40 years of bonus time. Because I was still active and otherwise healthy, and because surgery offered the promise of a complete cure, the decision to proceed was obvious. Many other patients in the 40-to-70 range are in a similar situation. If a simple surgical procedure can clear an obstructed artery, or medication can control your high blood pressure, and you can resume living your normal life afterward, you agree to the treatment without hesitating.

But that pattern can establish unrealistic expectations. A 70-year-old with congestive heart failure goes to the doctor anticipating the same outcome as a 40-year-old with high cholesterol: *Can you fix this, please? And can you make it quick? I've got a lot going on.*

But as the patient approaches 80, the symptoms of disease and decline become more pronounced. So do the side effects of any prospective cure. Eventually both the patient and the doctors have to confront hard questions. When does life stop being life and instead become a prolonged exercise in not dying? What should you do when you reach that point? And who decides?

~

And it's not as if medical science is going to stop now. New fields like genome editing are just dawning, amid predictions of revolutionary results. "Deleting genes could boost lifespan

113

by 60 percent, say scientists," a recent headline in *The Telegraph* read. That would mean people could routinely live to 128, or six years longer than the longest human lifespan documented so far.

First of all, how many people would really want to live that long? I'm not even halfway to 128 yet, and I can't imagine where I would get the energy (not to mention the money) to plug along for another 70 years. Nor can I imagine what kind of shape I would be in from, say, 112 on. Even when human life expectancy was only 40 years, 80-year-olds were not uncommon. It's just that the mortality rate across all ages, especially newborns, was so high that it brought the average lifespan way down. But because there are no documented 128-year-olds alive anywhere today, we have no idea what life would look like at that age. My hunch is that it would be far uglier than 80 is today. Especially when you consider how overweight and unhealthy many modern Americans are. Life expectancy in the U.S. has actually dropped in recent years. Picture generations of increasingly obese and unhealthy Americans sustained by artificial means far beyond their organic lifespans.

In other words: Genome editing (or some other marvel of modern science) won't cure the problem of a protracted and ambiguous end-of-life scenario. It will simply defer it. And possibly make it worse.

The added burden on society would be enormous. The number of centenarians in the U.S. has grown by almost 50% since 2000. Imagine if it were to grow by 500%, or 5,000%. Think of the strain that would put on the country's resources and its economy. If Social Security and Medicare are on the brink of insolvency now, what happens when people want to retire at 70 and collect benefits for decades on end? And when Medicare inevitably collapses, who pays for the end-of-life propping-up for all those new centenarians?

Beyond raising many practical and ethical questions, chronic end-of-life care further distances Americans from reality. Death has been divorced from life. "Dying in one's own bed is a rare privilege, an outlier in the calculus of modern dying," Dr. Warraich writes. "Not too long ago, death was not something that happened in hospitals or cloistered facilities; it happened in real neighborhoods surrounded by real people."

Now, instead of definitively dying, many Americans slowly disappear from life. Aunt So-and-So is forced to leave her home and go to a facility. By the time she actually dies a year or two later, most of those who knew her have already accepted her absence. Death is just the confirmation of something they've expected for a while. They feel no surprise, just relief tinged with melancholy.

For Aunt So-and-So, the long, drawn-out death in a lonely institution is miserable. But for those she's left behind, the gradual transition softens the blow.

Maybe that's why deaths that are sudden and unexpected, such as those that result from accidents or natural disasters, spark such intense emotional responses. In many cases, those close to the victims experience something beyond overwhelming grief. They're consumed with outrage. In the First World, death is no longer an acceptable part of life.

CHAPTER SEVEN

Recalculating

"Will you live at your own pace?"
—System of a Down

The most challenging aspect of living in a technologically interconnected society is trying to sync your core beliefs, and your rhythms, with everyone else's. You're constantly colliding with people who are on radically different trajectories, all inhabiting the same multilevel space.

You can choose *not* to interact with people through all that technological interconnectedness, of course. But that's an extreme choice that has profound implications not only for you but also for those close to you. (And a lot fewer people will remain close to you if you withdraw, or choose not to participate to begin with.)

What the hell am I talking about? Well, just ask any older person who never learned to text if they feel increasingly out of the loop, even with immediate family.

I'll offer my experience as a parent as another example. When our son was born, my wife and I vowed to shield him from junk food and junk media. But it didn't take long to realize that raising him like that in America's middle class would be tantamount to raising him in isolation.

Our son is a digital native. He was born in 2007, the year the iPhone debuted. Neither my wife nor I have ever owned an iPhone. We get by with cheap phones and prefer other devices for work and social interaction (I have a laptop; my wife favors an iPad). But most of our friends and relatives are more technologically progressive than we are, so it's not as if our son has never seen an iPhone. In fact, he's had enough exposure to iPhones to wonder aloud why we don't have them.

In other words, a normal middle-class American upbringing has conditioned our son to think that interacting with a device is an integral part of life. And, naturally, he wants what he thinks is the best device available. It's only a matter of time until he has one. (If it comes down to it, he'll buy one himself when he's old enough.) Short of keeping him away from digital devices altogether—which is a severe social handicap today—I don't see how we could have avoided that outcome.

As for preventing him from eating junk food: Almost from the moment my son graduated from formula to solid food, his diet has included pizza and ice cream—because that's the standard menu at any organized gathering for kids. Pizza is also a staple on my son's school lunch menu, along with cheeseburgers, chicken fingers, mac and cheese, spaghetti and meatballs, fish sticks, French toast sticks, tacos, and so on.

My wife and I could have shielded our son from exposure to industrially processed food by refusing birthday party invitations, by issuing strict instructions to doting relatives, and by sending him to school with veggie wraps and carrot sticks instead of lunch money. But that would have made him feel singled-out and self-conscious, at a time when developing an ability to assimilate with his peers is critical. So we opted to let him learn the value of breaking bread with people, even if that bread isn't very good for him. Had we monitored his diet too strictly, I have no doubt that by now he would feel deprived and socially awkward. I think the same would be true if we had monitored his media diet too strictly. (He was about four when one of his older cousins showed him the Black Eyed Peas' "Boom Boom Pow" video. After that he wanted to keep up with popular music and lost interest in insipid childhood standards like "Baby Beluga.")

When our son started fifth grade, my wife and I wondered after a few weeks why he wasn't bringing home corrected papers for us to see, as in years past. Eventually we learned from his teacher that his work was now posted online instead. For me, that was another *Well, duh* moment. Of course elementary schools are going paperless. So are banks and medical offices and virtually every other American institution.

I don't have any problem with elementary schools having kids do their work on screens instead of on paper. I think it's good, actually. I don't see any inherent value in handwritten schoolwork. When my son was learning cursive writing, I wondered why the school even bothered. Cursive writing feels like a quaint artifact—like the hole in my first-grade desk where an inkwell used to be. I can't imagine my son will use cursive for any purpose other than signing his name. That's all I ever use it for now. And scrawling my signature is no longer an autonomic act, particularly on electronic pads. My hand jitters like the needle on a polygraph.

Maybe that's just a result of fading muscle memory. Or maybe it's a sign of psychological distress, a subconscious questioning of my identity. But, really, I think what I'm wrestling with is less an identity crisis than a reality crisis. I feel pretty good about who *I* am; it's the world I have doubts about. Most of the time I'm consumed with the logistics of life in a small New Hampshire town. Work deadlines, Little League games, grocery lists, monthly bills, family gatherings, yard work, my son's math homework, dentist appointments, and so on. Life as a to-do list. It's the life I've chosen, and it's the life I want.

At the same time, I can't ignore the darker reality that lurks just beyond the warm confines of my day-to-day life. Through the window of the office where I'm writing this I can see the remnants of an old stone wall that a farmer put up ages ago, in a different world. I try to picture a mushroom cloud rising on the horizon beyond that old farmer's wall. The vision is unimaginable but not unthinkable. Not with almost 16,000 nuclear weapons in the world and people like Donald Trump and Vladimir Putin and Kim Jong-un calling the shots. No setting, no matter how idyllic, is safely beyond reach.

Or I think about the two once-a-century nor'easters that recently battered the New England coast within a month. I think about the Gulf of Maine, where temperatures are rising more rapidly than just about anyplace else on Earth. I picture a massive hurricane holding together long enough in the warming waters to make landfall in New Hampshire. I picture it swamping the Seabrook Station Nuclear Power Plant—which was built with a startling lack of foresight in a saltmarsh right on the coast.

The Nuclear Regulatory Commission recently extended Seabrook's operating license until 2050, despite some projections that the sea level along New Hampshire's coast could rise by as much as 21 inches by then; despite the absence of a plan for removing spent fuel rods for the foreseeable future, which means Seabrook will continue to store them onsite, in concrete casks; and despite the presence of something called alkali silica reaction, which has already reduced some concrete structures at Seabrook to a condition that regulators call "operable but degraded and nonconforming."

That all sounds to me like an engraved invitation to a Fukushima-style meltdown that could turn the entire New Hampshire coastline into an exclusion zone. I picture honkytonk Hampton Beach abandoned like that haunting amusement part near Chernobyl.

Or I ruminate about yet another potential doomsday scenario that's been drawing increasing attention: "Catastrophically Dangerous AI," as one headline described it.

It doesn't take much imagination to picture what could go wrong if smart machines eventually get smarter than humans. Documentary filmmaker James Barrat explored that possibility in a book called *Our Final Invention: Artificial Intelligence and the End of the Human Era*. He concluded that "AI is a dual-use technology like nuclear fission. Nuclear fission can illuminate cities or incinerate them. Its terrible power was unimaginable to most people born before 1945. With advanced AI, we're in the 1930s right now. We're unlikely to survive an introduction as abrupt as nuclear fission's."

Barrat's fear—a fear that a growing number of scientists share—is that AI will morph into ASI, or artificial super-intelligence, which will no longer be constrained by human oversight. And if humans recklessly unleash ASI to try to solve all of the world's problems, ASI might conclude that humans are the biggest problem of all. ASI might then decide to eliminate that problem.

~

Writing a book about such weighty concerns feels like a big responsibility. I feel like I should try to outline solutions to the problems I've described at such length. But I'll be the first to admit that I don't have all the answers. In fact, I don't have any of the answers. All I have is an endless string of questions. *How did the world come to this? Why do we accept it? How can we get Americans to shake off their self-absorption and start taking these problems seriously?*

Ultimately the purpose behind a book like this is to keep repeating the questions and rephrasing them, like a lawyer grilling someone on the stand, hoping to shed new light, provide fresh insights, and offer a different perspective.

So, in that spirit, I'd like to call my final witness.

Earlier in this book, I imagined what it would be like to take a day trip to the 1800s. Now I'm going to complete that exchange program. I'm going to imagine what it would be like to invite someone from the 1800s to spend a few days at my home in New Hampshire.

I'll call him John. That was the most common name given to males born in the 1830s (1836, in John's case). He's 46 years old. He's joining the present from the year 1882, which means that he knows about the telephone, photography and electric light. He can't wait to see how much more technological progress the human race has made by the 21st century. And because he's bright and has a lot of curiosity, particularly about science and engineering, he'll be able to make the intellectual leaps required to understand many of the devices he'll see in the future—though certainly not all of them. (I couldn't begin to explain Google Glass to him.)

I wouldn't reach any further back in time for this exercise. I think anyone who lived before the 19th century would simply suffer a breakdown if suddenly transported to the 21st century—even someone as visionary as Leonardo da Vinci or Benjamin Franklin.

One more thing about John. He remembers the horror of the Civil War. He served in the Union Army. He saw things on the battlefield that still haunt him.

~

So, here's how it went.

It took John a full day just to get acclimated to the house. And I didn't even show him the TV, internet, or car right away. I wanted to work up to those.

John was fascinated enough by the house itself. First, he couldn't believe that my wife, son, and I had such a big, comfortable space—and that we weren't rich or privileged.

Nor could he understand how the house could stay warm and free of drafts on a cold, breezy day. (It was early spring, with a little snow still on the ground.) "Where's the fire?" he asked.

I showed him the boiler in the basement and the heating-oil tank and the pipes for the baseboard heaters. I explained about insulation and energy-efficient windows. (He studied the windows far more closely than I ever have and asked several questions that I couldn't answer.)

The bathroom also amazed him. He couldn't get over the combination of indoor plumbing and hot water that we all take for granted. (I didn't think he would ever come out of the shower.) Everything was "stupendous!"

John loved our kitchen, even though it's nothing special and most of our appliances are out of date. He had a bunch of questions about how the refrigerator worked. Again, I couldn't answer most of them very thoroughly. (I know refrigerators used to use Freon, but I don't know what they use as a refrigerant now. I started to explain about fluorocarbons and ozone depletion but thought better of it. No sense freaking him out so early in his visit.)

To John, the easy availability of so much fresh food was a minor miracle—as was the ability to cook it so quickly using the electric stove or propane grill or microwave (the workings of which, like energy-efficient windows and the refrigerator, I couldn't explain very well).

I'd planned a careful menu for John's visit. I rounded up as much simple, basic food as I could: fresh bread from our local bakery (instead of the store-bought stuff), locally grown eggs, a basic roast beef with organic vegetables including potatoes and carrots. I figured that would come closest to what he was used to. But I also wanted to get his reaction to such modern staples as cheeseburgers and pizza.

These were his general impressions:

Starbucks coffee: "Very good."

Scrambled eggs with buttered toast: "Very good."

A frosted cherry Pop-Tart: "Too sweet." (I think he had cavities and the sugar made his teeth hurt.)

Roast beef with potatoes, carrots, and gravy: "A bit bland, but satisfying enough."

Pepperoni pizza: "Peculiar, but tasty."

A cheeseburger with ketchup: "Quite good."

Doritos: "Dreadful! Vomitous! Perhaps I'll try another."

Most surprising was John's reaction to craft beer. He liked it. I gave him an IPA with a hint of grapefruit, which he said reminded him of something he once tried in New Jersey called "California pop beer." (I Googled it and found that it contained spruce oil, gingerroot, sassafras, and even wintergreen.)

John quickly grew bored with my taste tests. He was astute enough to recognize that his opinion of any particular food was beside the point. He grasped the profound change that refrigeration and prepared foods had wrought in people's lives. He marveled at the autonomy 21st century Americans had when it came to their diets, not to mention how little time they devoted to acquiring and preparing food, relative to his era. He couldn't see any downside.

~

I introduced John to electronic media by degrees. I showed him email first. I figured that was the closest analog to the technology he was used to. Basically email is a telegram that appears as text on an electronic screen (which, of course, he spent considerable time marveling over and asking questions about—questions that further underscored my ignorance).

From email it was a fairly short leap to an online newspaper. John got it right away. If an individual could send text through the ether, why couldn't a newspaper do the same? He was surprised to learn that both of those developments had happened fairly recently. And he was astonished when I told him that manned flights to the moon had predated commonly available email. (Also, he was thrilled when I told him there was a reasonable chance that, back in his own era, he would live to see people fly.)

The first video I showed him was *A Trip Down Market Street*, the short film made in San Francisco in 1906. It was hard for me to believe that it looked futuristic to him. The horseless carriages blew him away. "Wait till you see what they look like now," I told him.

After that I showed him Neil Armstrong's "One giant leap for mankind" moment. His lip quivered as he watched it. And for once I was actually able to answer most of his questions. We took a break after that.

Finally, I turned on the TV. As you might guess, John was almost overwhelmed by the technology. And I was back to not knowing the answers to most of his questions. He was much more fascinated by how things work than most 21st century Americans are. Our only concern is *if* things work.

Although he was enthralled with the technology, John had little interest in the content of anything on TV or online. He couldn't follow it. That's understandable; most of what people were talking about meant nothing to him.

Also, he thought the names of many modern applications, like Google and Twitter and YouTube, sounded foolish. "This all seems rather juvenile," he said. I laughed.

He found TV too loud and frenetic. The editing was too fast and the music harsh and unfamiliar, using mostly instruments that he didn't recognize. Commercials in particular threw him off. He would be trying to follow the narrative thread of a program when it would jump to a commercial without warning. He quickly tired of the TV and asked me to turn it off.

He had a similar response when I took him for a drive. As expected, he was fascinated with the car at first—even though he had seen modern cars on TV, which spoiled my intended surprise.

Our car was just a Subaru Outback with 100,000 miles on it. Even so, it was the most sophisticated mechanical device he had ever seen. He wanted to know everything: what it used for fuel, how the starter worked, how the tires managed to hold air, how the all-wheel drive functioned, and on and on. Again, my ignorance about the underlying technology that enabled much of my way of life was embarrassing. I could see John getting frustrated at times. ("Were you not ever curious enough to ask that question yourself?") We talked about the car for more than an hour before we actually went anywhere.

I started out on the back roads near our house so John wouldn't feel overwhelmed. He had ridden plenty of trains, so the speed didn't bother him. He was amazed at how quiet and smooth the ride was, though. He was stunned when I told him the entire country was now crosshatched with paved roads, and that we could go all the way to California without ever once driving on dirt. "And every family has one of these machines?" he said.

"Actually, most families have more than one."

He was silent for a bit. I think he was considering the implications. To a 19th century man, the idea that virtually every person in America could go anywhere in the continental U.S. at any time was difficult to comprehend.

John flinched the first time a car came at us from the other direction. I explained that everyone understands that they need to stay on the right side of the road. He found it hard to believe that every person with a car would actually stick to that rule.

He had an even more pronounced skepticism about traffic signals. The first time I came to an intersection with a light, he panicked when he realized I wasn't going to stop. He could see another car approaching from the right. He yelled and put his arms up when I proceeded through the intersection, right in the other car's path.

"I knew that other driver would stop, because he had a red light," I said. "Did you notice that our light was green? That's the universal system for traffic signals. Green always means go and red always means stop. There's also a yellow light that serves as a warning between a green light and a red light, so you have you time to stop—or to speed up and get through before it changes, which is what most people try to do."

John thought about that for a moment. "How did you know his light was red?" he said.

Good question. I told John that a green light for one direction always matched up with a red light for cross traffic. But it was really just a leap of faith. Drivers actually don't know what color the signal is for cross traffic. It's a little disconcerting when you realize how much trust we place in traffic lights, not to mention every other driver we cross paths with, whenever we venture out of the road.

Eventually I came to the main commercial intersection in town, where all the franchise restaurants are clustered. There were four lanes of traffic in each direction—two that went straight and two that turned left, as indicated by green arrows.

That did it. John's anxiety was palpable now. He asked me to take him back to the house.

~

When we got back from the drive, John asked for another IPA. I was happy to join him.

He grew pensive. "The most heartening thing about this visit," he said, welling up, "is learning that the Union survived." He told me that his greatest fear was that the rift caused by the Civil War—or "the great rebellion," as he called it— would never fully heal. He worried that the U.S. would eventually disintegrate and his descendants wouldn't have a permanent home.

I told him that the U.S. had not only survived, it had thrived. During the 20th century it had become the world's leading power, and winning the race to put the first man on the moon typified what many people worldwide thought of as the American spirit.

He teared up again. "And seeing such wealth and comfort and contentment among so many citizens such as yourself—may I presume that our great nation is at last at peace?"

I paused before answering. "No, not entirely," I said. I briefly explained about our involvement in Afghanistan. That boggled his mind. I could read the expression on his face: *Afghanistan? What possible quarrel could we have with Afghanistan?*

Then I told him that the greatest threat to the United States was kind of faceless and abstract. It lay in the collective power of the world's weaponry, which had expanded exponentially beyond what was available in 1882.

"So there are great stockpiles of weapons everywhere?" he said, frowning.

"No—well, yes, actually," I said. "But the real problem is how powerful certain weapons are. They're literally beyond anything you can imagine."

His frown deepened and he was silent for a bit. Then he drew a deep breath. "I was at Gettysburg," he said, his voice catching. "I don't have to imagine the horror of war. I lived through it. I can't fathom how anything could be more terrible than what I observed during those three days."

"I respect that," I said. "But what I'm talking about is a matter of amplification. There are weapons today called thermonuclear bombs that could have killed every single soldier at Gettysburg, on both sides, within a matter of seconds."

I could see John struggling to understand. "How could that be?" he said. "Even if I accept that such powerful bombs now exist, to use them would be suicidal, would it not? What would be the tactical advantage in killing all of your own troops as well as the enemy's? Such weapons would be of little practical use on the battlefield."

"That's correct," I said. "Thermonuclear weapons would be of little use on a traditional battlefield. But they're not designed to be used on a battlefield. They're designed to be used from a great distance. You deploy them using what's called a missile—which is similar to the rocket that delivered men to the moon. You keep them armed and ready to fire at all times, and you store them in underground silos that are like caves. Many of ours are out in what you would call the Dakota Territory."

"To prevent Indian attacks?" John said.

I almost laughed but I managed to hold it in. "No," I said. "Indian attacks are no longer a problem. All the Indians are essentially gone now. I mean, they're still here, but ... it's complicated. The point is, Indians aren't the reason for our system of nuclear defense."

"So who are the ... what are the weapons called again?"

"Thermonuclear missiles."

"So who are the thermonuclear missiles protecting us against? Have the British reclaimed Canada?"

"No, Canada is one of our allies, along with all of Great Britain. Still, we have quite a few enemies around the world— although in most cases they are enemies only in the sense that they have a different form of government than we do as opposed to being in active conflict. But the ever-present fear is that that could change at any moment, for reasons that are difficult to foresee."

I took a sip of beer. "Anyway," I said, "the short, oversimplified answer is that our thermonuclear arsenal protects us from Russia and China."

"Russia and China!" John said. "Why, then, are we keeping the missiles in the Dakota Territory?"

"Well, we have missiles in other strategic locations around the world, too. But our ICBMs—that stands for 'intercontinental ballistic missiles'—can fly at speeds of four miles per second and can reach targets more than 8,000 miles away. You can fire a thermonuclear missile from the Dakota Territory and it will reach Russia in about 30 minutes."

John was silent again, his face a mask of worry and bewilderment. "But how do you aim those missiles?" he said at last. "How would you know where the Russian army is at a particular moment?"

"It doesn't matter," I said. "Our missiles aren't aimed at a particular army. They're aimed at cities—Moscow in particular, because that's now Russia's capital instead of St. Petersburg. Thermonuclear bombs can annihilate entire cities in an instant and in all likelihood would eliminate the enemy government. Imagine if the Confederates could have detonated a bomb powerful enough to have destroyed all of Washington and killed not only President Lincoln but also all of his cabinet along with our congress. The Union would suddenly have been without any leadership, and the loss of our capital would have been a devastating blow. The Confederates would have won the war. Unless, of course, we had been able to annihilate Richmond at the same time they annihilated Washington. Then both sides would have degenerated into chaos and anarchy.

"And that would be the outcome of each side firing just *one* thermonuclear missile. If a full-scale nuclear exchange ever happened, there would be thousands of missiles criss-crossing in the skies all at once. Basically the entire world would become Gettysburg in an instant, possibly without any warning. Thousands of places across the United States would be potential targets, including some near where we're sitting right now."

I noticed that John's fingers were interlaced and his knuckles were white.

"And honestly, those of us closest to the intended targets would probably be the lucky ones because we might not ever know what hit us," I said. "Survivors in outlying areas would suffer in unimaginable ways. First, there would be the incalculable trauma of seeing the modern world reduced to a primitive state in a matter of hours. Second, no one can really predict what the physical outcome of a widespread nuclear conflict would be. In addition to the massive blasts, thermonuclear bombs also release large amounts of poison because of the materials they're made with. Also, the enormous clouds of dust in the air could dramatically cool the planet and lead to crop failures and widespread famine."

John put up his hands and closed his eyes. "No more!" he said.

We sat in silence for a while. When John finally spoke, his voice was barely audible. He asked how long nuclear weapons had been in existence. "Since the 1940s," I said. I explained about World War II and the Manhattan Project. I showed him YouTube clips of the aftermath at Hiroshima and Nagasaki and gave him a crash course in the nuclear arms buildup that followed. John was deeply troubled that the U.S. was the only nation ever to have used nuclear weapons in war.

He also found it difficult to believe that so much time had elapsed without another episode of nuclear war. "You have lived your entire life with this threat, this fear?" he said.

"Yes," I said. "And oddly enough, I don't remember the first time I ever heard someone talk about nuclear weapons or understood the existential threat they posed. It just seems like I've always known about them, in the same way that kids gradually come to realize that they're going to die someday. And your reaction is similar. You don't dwell on the possibility—you just go on with life as best you can."

John's distress was visible. His lips were pursed and his brow was deeply furrowed. "I do not think I could ever achieve such a state of equanimity," he said. "In fact, I would like to return now to my own time. And I would like to have all memory of this visit eradicated."

"I understand," I said.

We stood to say our good-byes. When I extended my right hand, John gripped it firmly in both of his. "May God have mercy on you all," he said. He took a last look around my house. "Your many comforts are a thing of wonder. But I fear that all of your prosperity and the stupendous mechanical wizardry that you enjoy has come at a frightful price."

"No argument here," I said.

Made in the USA
Columbia, SC
22 October 2020